# Engineering Materials

This series provides topical information on innovative, structural and functional materials and composites with applications in optical, electrical, mechanical, civil, aeronautical, medical, bio- and nano-engineering. The individual volumes are complete, comprehensive monographs covering the structure, properties, manufacturing process and applications of these materials. This multidisciplinary series is devoted to professionals, students and all those interested in the latest developments in the Materials Science field, that look for a carefully selected collection of high quality review articles on their respective field of expertise.

**Indexed at Compendex (2021)**

Jirut Meesane

# Mimicked Tissue Engineering Scaffolds for Maxillofacial and Articular Cartilage Surgery

Springer

Jirut Meesane
Division of Biomedical Science
and Biomedical Engineering
Faculty of Medicine
Prince of Songkla University
Songkhla, Thailand

ISSN 1612-1317 ISSN 1868-1212 (electronic)
Engineering Materials
ISBN 978-981-19-7829-6 ISBN 978-981-19-7830-2 (eBook)
https://doi.org/10.1007/978-981-19-7830-2



This Springer imprint is published by the registered company Springer Nature Singapore Pte Ltd.
The registered company address is: 152 Beach Road, #21-01/04 Gateway East, Singapore 189721, Singapore

# Contents

# About the Author

**Dr. Jirut Meesane** worked in the group of Professor Dr. Karl-Friedrich Arndt, Institute of Physical Chemistry and Electrochemistry, Physical Chemistry of Polymer, Faculty of Natural Science, Dresden University of Technology, Dresden, Germany, 2004. Then, he started to work as Dr-Ing candidate in the group of Prof. Dr. Wolfgang Pompe and Dr. Michael Gelinsky, Max Bergmann Center of Biomaterials, Institute of Materials Science, Faculty of Mechanical Engineering, Dresden University of Technology, Dresden, Germany, 2005. His research work was focused on tissue engineering scaffolds. He obtained Dr.-Ing from Technical University of Dresden (TU Dresden), Germany, 2009. Then, he jointed to Faculty of Agro-Industry, Prince of Songkla University, as Lecturer in the Department of Materials Product Technology. In 2011, he has moved to Institute of Biomedical Engineering, Faculty of Medicine, Prince of Songkla University, as Lecturer and Investigator in biomaterials, biomimetic, and tissue engineering until now. In 2011 and 2016, he was promoted as Assistant Professor and Associate Professor in Institute of Biomedical Engineering, Faculty of Medicine, Prince of Songkla University, respectively. His works have been emphasized on tissue engineering scaffolds based on mimicking design for surgery. Those scaffolds were mainly focused on oral, cranio-facial, and orthopedic surgery. Furthermore, he has started to carry on the scaffolds with multifunction for tissue disease treatment. He has also developed the created scaffolds with specific function as biomaterials for diagnostic and evaluation of biological function of bioactive molecule. He has both national and international research collaborations with scientist, engineer, and clinician in university and research institute.

# Chapter 1
# Introduction

Tissue engineering is an attractive approach to induce new tissue formations at a defect site, due to disease or trauma [1]. In many cases, tissue engineering has been proposed as the method to use in surgeries, particularly in maxillofacial and articular cartilage applications [2, 3]. Performance scaffolds are especially used as biomaterials for surgery to enhance tissue formation at defect sites [4]. To create performance scaffolds, mimicking is often selected as the approach in their design to promote their potential for enhancement of tissue formations [5]. Included in this chapter are main three parts: 1) principles of tissue engineering, 2) mimicking of tissue engineering, which is focused on structure and function in scaffolds, and 3) mimicked scaffolds in maxillofacial and articular cartilage surgery.

## 1.1 Tissue Engineering

Tissue engineering is a performance-based approach, which has been used to repair defects, based on new tissue formations. Basically, tissue engineering has three main parts: cells, growth factors, and scaffolds. These three main parts are important in the designing of a system to regenerate new tissue. The scaffolds themselves exist as a home, which is decorated with growth factors. Growth factors demonstrate unique functionalities to induce cell behavior, namely growth, proliferation, migration, and differentiation inside the scaffolds. Therefore, fabricating suitable scaffolds which have been modified with effective growth factors is important in the design of a system for new tissue regeneration. In this book, the subject of tissue engineering scaffolds is the main topic. The scenario of tissue engineering is shown in Fig. 1.1.



J. Meesane, *Mimicked Tissue Engineering Scaffolds for Maxillofacial and Articular Cartilage Surgery*, Engineering Materials,
https://doi.org/10.1007/978-981-19-7830-2_1

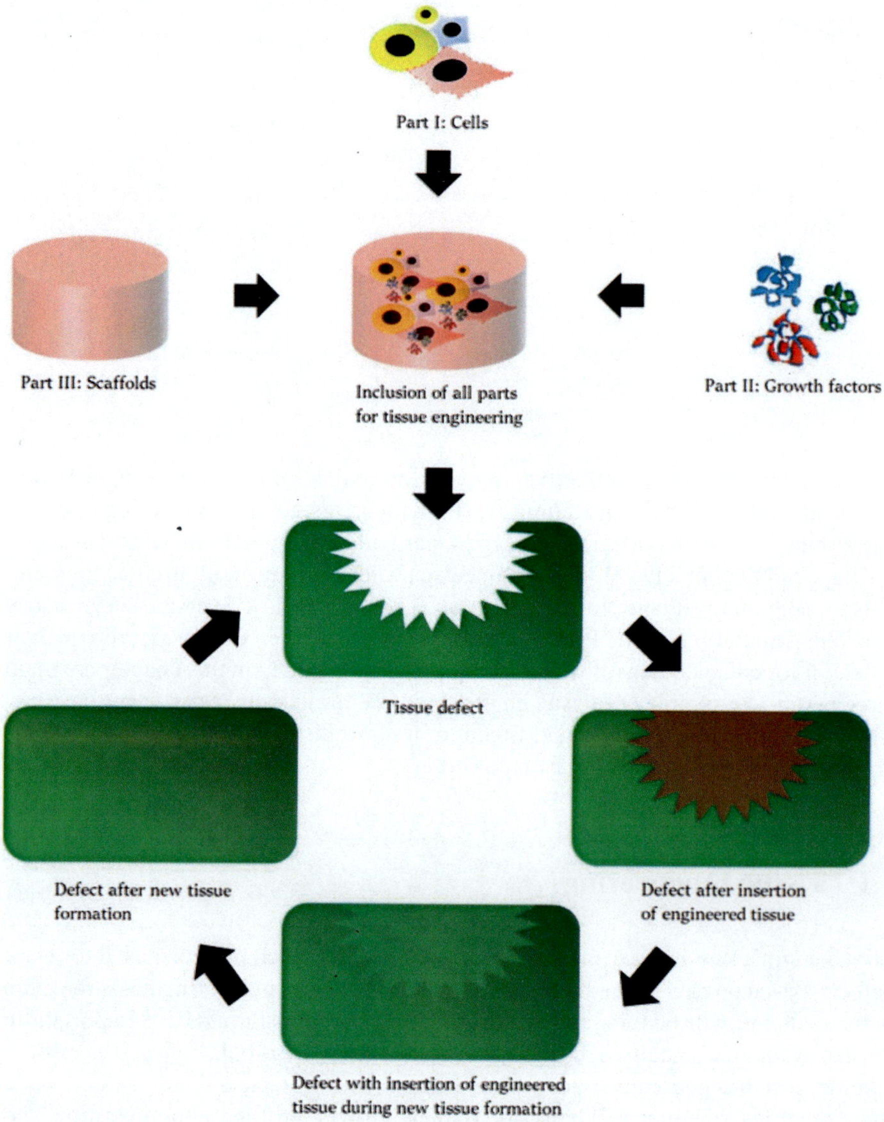


**Fig. 1.1** Scenario of tissue engineering

### *1.1.1 Cells*

Cells are the main components of tissue. They show the behaviors of adhesion, proliferation, migration and differentiation, which in turn leads to tissue formation [6]. Cells are important for the creation of tissue engineering systems [7]. For application in tissue engineering, cells are added into the defect site, to induce the new tissue formations [8]. There are many approaches to create the potential of cell systems for tissue engineering. Both stem cells and primary cells are often used to promote tissue formation, especially stem cells, which have the potential to induce tissue formations [9]. For some cases, stem cells are co-cultured with the other cells to engineer tissue, which is similar to natural tissue [10]. Furthermore, to complete tissue engineering, cells are often used with growth factors and scaffolds. For this book, cells are simply described to design suitable systems for tissue engineering.

### 1.1.2 Growth Factors

Growth factors are biological signals that have the role of inducing cell behaviors: adhesion, proliferation, migration, and differentiation, which lead to the induction of new tissue formations [11, 12]. In the interesting applications of tissue engineering, growth factors releasing are used to promote tissue formation. Some designs demonstrate dual releasing, which is created as programming to regulate tissue formation, while some systems are created for multi-releasing of growth factors, so as to stimulate and regulate tissue formations. Hence, understanding the functional sequences of growth factors throughout new tissue formation is important in the design of those programs [13, 14]. Furthermore, to create programed releasing of growth factors, it is important to design potential systems to induce tissue formation [15]. In this book, growth factors are briefly explained to design the system for tissue engineering.

### 1.1.3 Extracellular Matrix

The extracellular matrix (ECM) is a natural scaffold, which is the performance of biomaterials for tissue engineering [16]. ECMs have the important role of serving as residents for cells [16]. In the construction of a tissue engineering system, cells are seeded into the scaffolds. The cells are then cultured into a new tissue formation. The scaffolds need to perform physically, mechanically and biologically in such a way as to enhance appropriate cell behavior, and thus inducing new tissue formation [17, 18]. In the selection of materials, the important requirements of scaffolds are biocompatibility, biodegradability, and non-toxicity [19]. After selection of the proper materials, a suitable structure coupled with functionality need to be designed for a tissue engineering system [20, 21]. In this book, the structure and function of ECM, as the model to design scaffolds, are considered and discussed. In this book, ECM is mainly used as a model in the guidance to create performance scaffolds for tissue engineering.

## 1.2 Mimicking for Tissue Engineering

Mimicking is an interesting method, which has been used in the design of many innovative applications, such as structural and mechanical engineering, along with architecture [22, 23]. For tissue engineering, mimicking has been used to create performance systems to regenerate new tissue [24]. In the view of tissue engineering, mimicking has been a guide in the design of the cells, growth factors, and scaffolds [25]. Especially, scaffolds have mimicked both structure and functionality similar to that of natural tissue. This being the main focus of this book, mimicked structures initially focus on the scale, from a molecular to a geometrical structure. Next, mimicking of functionality is considered in physical, mechanical, and biological performance. These three main parts are often used for tissue engineering scaffolds. The concept of mimicking for tissue engineering is shown in Fig. 1.2.

### *1.2.1 Mimicked Structure in Scaffolds*

The structure of scaffolds is important to enhance their potential. The structure of scaffolds has been considered in three scales: (1) nano-scale, (2) micro-scale, and (3) macro-scale [26, 27]. For the nano-scale, molecules have been constructed into structures which are similar to the molecular structure of the extracellular matrix [28]. These constructions show the unique performance to self-organize into a structure similar to the extracellular matrix. A self-organized structure is able to enhance specific cell behaviors, which in turn can enhance new tissue formations [29]. At the micro-scale level, scaffolds have been fabricated into specific morphological structures, which are suitable for specific cell behaviors [30]. Porous and fibril structures are often considered as guidance to mimic these scaffolds [31]. At the macro-scale level, scaffolds have been constructed similar to the anatomical structure of natural tissue [32]. Because of the hierarchical structure, which has different types of tissue, biphasic scaffolds have been constructed to fit native tissue [33]. Different scales of structure are highlighted as guidance in the design of potential tissue engineering scaffolds.

### *1.2.2 Mimicked Function in Scaffolds*

The function of scaffolds has a significant effect on the behavior of the existing cells until complete new tissue formations [34]. The physical, mechanical, and biological functionalities have been often reported as guides to mimic the potential scaffolds [35]. The physical functionality of scaffolds has been modified with molecules, which could enhance their potential [36]. Hydrophilic molecules are often used to induce the swelling behavior, which can enhance nutrients. This leads to induction of

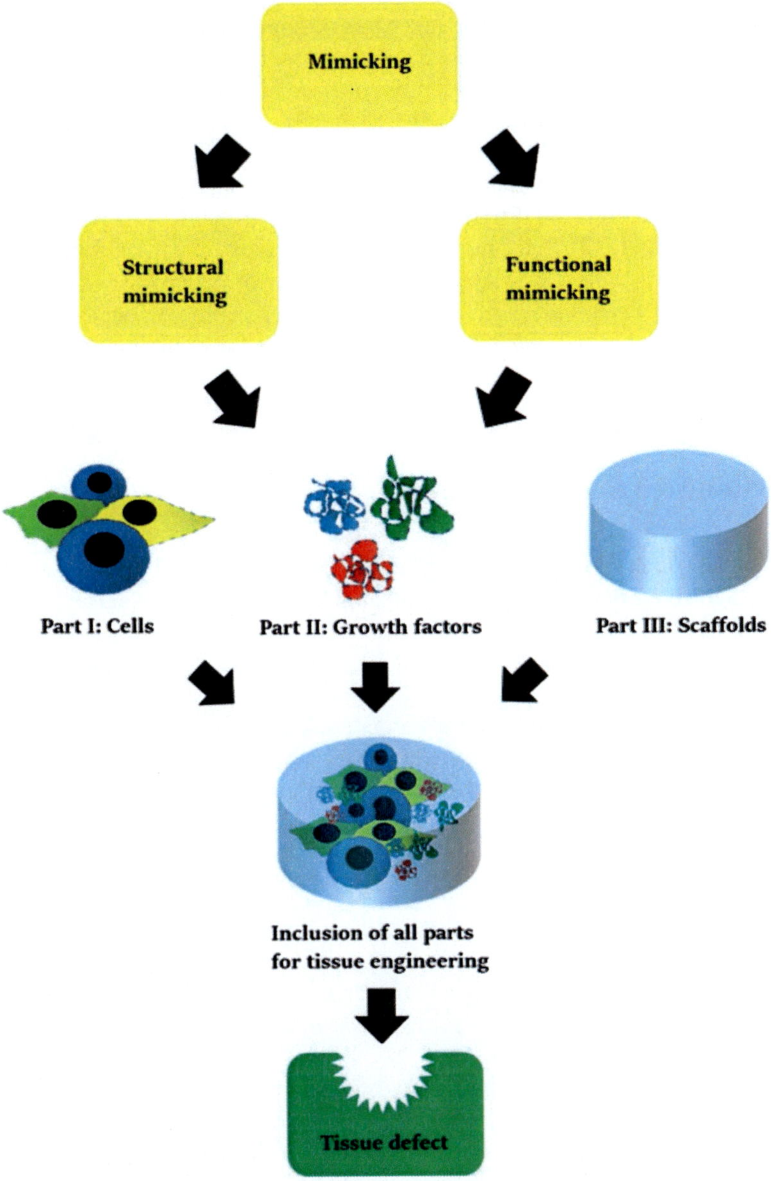


**Fig. 1.2** Concept of mimicking for tissue engineering scaffolds

new tissue formations [37, 38]. The mechanical functionality of scaffolds has been modified with crosslinking, which can induce structural stability [39]. Structural stability has an effect on the enhancement of mechanical strength, toughness, and stiffness. Mechanical modifications have been used as a guide to mimic scaffolds similar to natural tissue [40]. The biological functionality of scaffolds has been modified with molecules, which then act as recognized groups for cell adhesion, proliferation, migration, and differentiation. These have an effect on promoting new tissue formations [41, 42]. Some molecules have been grafted with scaffolds to induce specific protein adsorption, which enhances new tissue formation [43]. Mimicking for the functionality of scaffolds, similar to natural tissue is emphasized; especially, the mimicking of the physical, mechanical, and biological functionalities is considered and discussed.

## 1.3 Mimicked Scaffolds in Maxillofacial

Maxillofacial surgery is often used to treat patients who suffer from a serious tissue defect, either from a disease or from a trauma. There are many classical and advanced approaches to treat these patients, with biomaterial substitution being a popular one. [44, 45]. Scenario of the applications of biomaterials in maxillofacial surgery is shown in Fig. 1.3.

Classical approaches of long-term biomaterial implantation are still applied in some surgeries, especially in patients who have a severe case of a large defect area]. A large defect area can cause dysfunction of an organ. Performance biomaterial

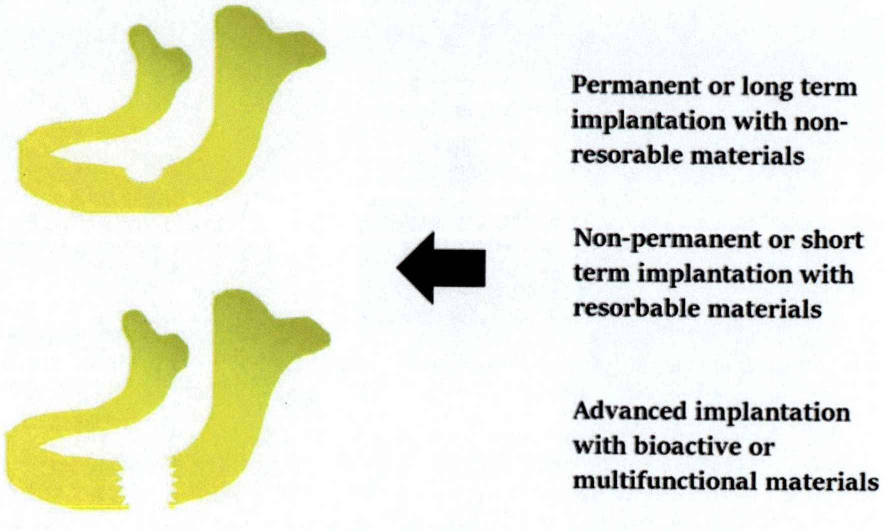


**Fig. 1.3** Scenario of biomaterials implantation in maxillofacial surgery

implantations have often been proposed for long-term implantation [46, 47] (Fig. 1.4). However, due to the aging effect, some defects on the biomaterials lead to a release of debris into the surrounding tissue, this can then lead to irregular formation of tissue around the biomaterial [48, 49]. In addition, some defects on the biomaterials lead to cracking, which in turn leads to non-fixation with the tissue, and calls for the requirement of a second operation [50].

To solve this problem, tissue engineering scaffolds, which are non-permanent biomaterials, are available for applications in maxillofacial surgery [51]. The scaffolds serve to maintain stability until new tissue is able to replace the whole defect site [4]. To create high-performance scaffolds, to induce tissue formation, they are often combined with cells and growth factors for application in surgery [52]. To succeed for promotion of tissue formation, scaffolds are created based on the mimicking of structures and functions similar to natural tissue [53, 54]. Furthermore, the mimicked scaffolds are required to have suitable function for surgery, for instance suitable handling during surgery and the ability for molding and fitting into the defect site. Hence, to create performance scaffolds, based on mimicking, is an important requirement in the perspective of tissue engineering and surgery. The reports of different

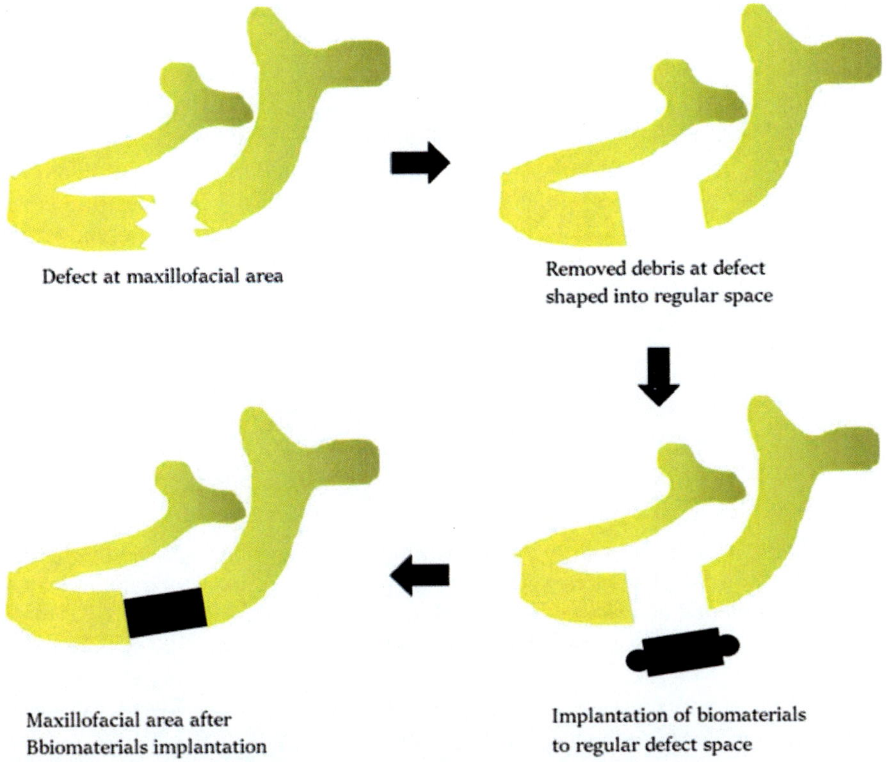


**Fig. 1.4** Scenario of permanent implant biomaterials in maxillofacial surgery

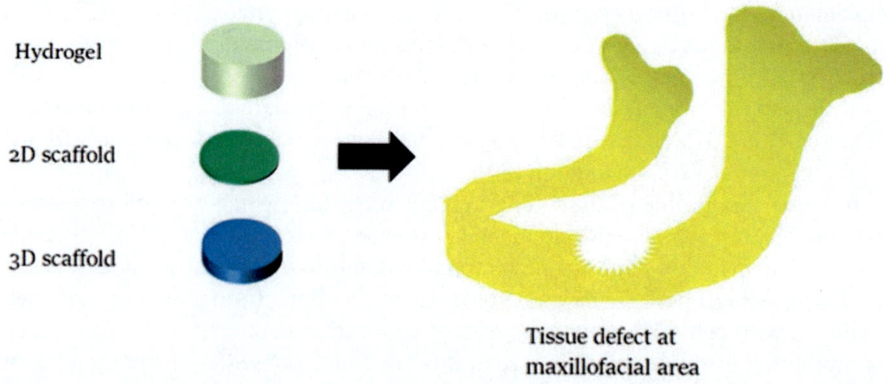


**Fig. 1.5** Tissue engineering scaffolds for maxillofacial surgery

types of scaffolds have led to suggestions and suitable selections for maxillofacial surgery. Hydrogels, two-dimensional (2D), and three-dimensional (3D), are used for maxillofacial surgery (Fig. 1.5) [55].

For maxillofacial surgery, mimicked scaffolds for bone tissue engineering have often been reported [56, 57]. They are then placed in these defects. It is important to mimic the structure similar of natural bone in the design of potential scaffolds for new tissue formations. After the selection of suitable materials, of either natural or synthetic polymers, scaffolds are fabricated into a stable structure that can maintain the contoured shape at the defect area [58, 59]. Crosslinking and reinforcement with bioactive molecules, particularly calcium phosphate, have been used in suitable hydrogel or scaffold modifications [60]. For biological modifications, polymers are often immobilized or mixed with biological signals, which can induce new tissue formations [61, 62]. Modified polymers are then fabricated into scaffolds [63, 64]. In other methods, scaffolds were fabricated into specific structures before modification [42, 65].

## 1.4 Mimicked Scaffolds in Articular Cartilage Surgery

For articular cartilage surgery, there are two main defects: (1) severe and (2) non-severe cases (Fig. 1.6) [66]. In severe cases, the patients need medical implant materials to substitute the defect area [67, 68]. Some cases need medical devices, which are designed to replace the knee joint (Fig. 1.7) [69]. This device needs long-term implantation.

In the case of non-severe cases in articular cartilage surgery, scaffolds are used to promote new tissue formations [70, 71]. These scaffolds have the main role as a structure for cell adhesion, spreading, proliferation, and migration, which lead to

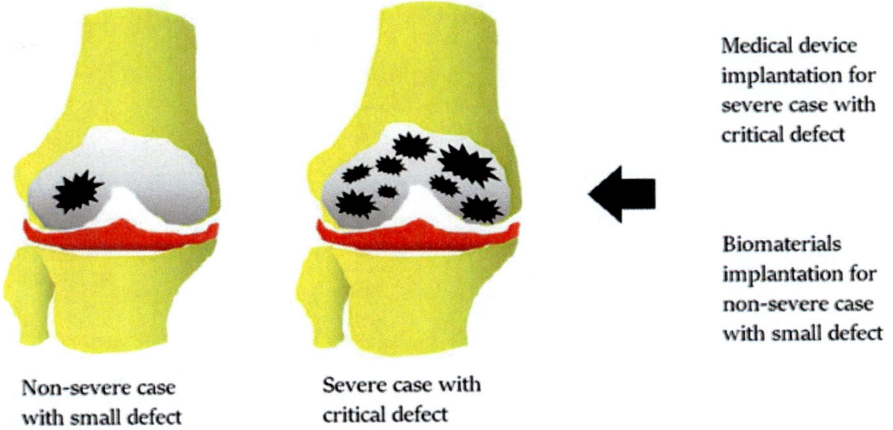


**Fig. 1.6** Defects at articular cartilage and proposed biomaterials implantation for articular cartilage surgery

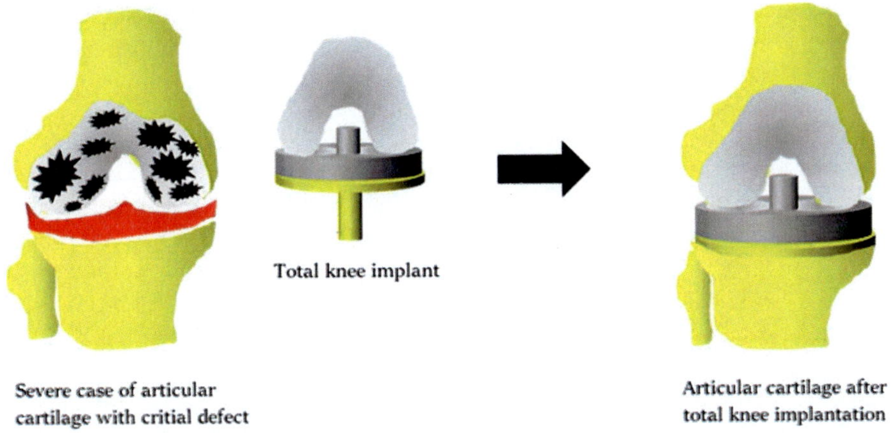


**Fig. 1.7** Total knee joint replacement for articular cartilage surgery

enhancement of new tissue formations [72]. The requirement of scaffolds is biocompatibility, biodegradation, and non-toxicity [73, 74]. Furthermore, they need suitable physical and mechanical properties to assist in maintaining their contoured shape until complete new tissue formation at the defect area has taken place [75–77]. The three types of scaffolds used for articular cartilage surgery are: (1) hydrogels and (2) two-dimensional (2D) and three-dimensional (3D) scaffolds (Fig. 1.8). To develop their performance, these scaffolds are often created with based mimicking. For instance, in the case of hydrogels, they are prepared into a highly viscous characteristic in order to have the physical performance similar to cartilage [78]. This physical performance has a high water content reservoir, which has an effect on

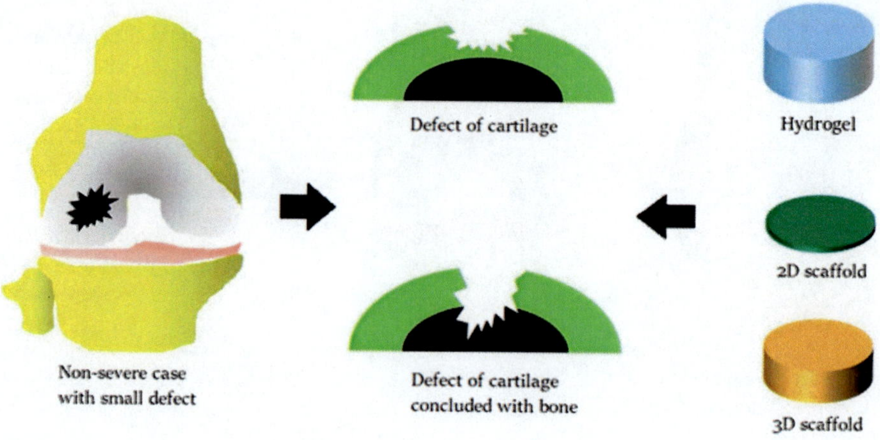


**Fig. 1.8** Scaffolds in articular cartilage surgery

the mechanical performance in the dispersion and resistance of forces during body movement [79, 80]. In the case of scaffolds, they are fabricated into mimicked structures and functionalities that show a suitable pore size for cell behavior that include adhesion, proliferation, migration, and differentiation [80].

Importantly, for the concept of this book, tissue engineering strategies are presented for use in maxillofacial and articular cartilage surgery. The focus is especially on tissue engineering scaffolds, which are created by the based mimicking. The principle of mimicking in tissue engineering scaffolds is discussed in more detail. Examples of mimicked scaffolds in maxillofacial and articular cartilage are also shown as guidance in the design of potential scaffolds.

## References


1. Shafiee, A., Atala, A.: Tissue engineering: toward a new era of medicine. Annu. Rev. Med. **68**, 29–40 (2017)
2. Sangkert, S., Meesane, J., Kamonmattayakul, S., Chai, W.L.: Modified silk fibroin scaffolds with collagen/decellularized pulp for bone tissue engineering in cleft palate: morphological structure and biofunctionalities. Mater Sci Eng C Mater Biol Appl. **58**, 1138–1149 (2016)
3. Thangprasert, A., Tansakul, C., Thuaksubun, N., Meesane, J.: Mimicked hybrid hydrogel based on gelatin/PVA for tissue engineering in subchondral bone interface for osteoarthritis surgery. Mater Design **183**, 108113 (2019)
4. Ceccarelli, G., Presta, R., Benedetti, L., De Angelis, M.G.C., Lupi, S.M., Baena, R.R.Y.: Emerging perspectives in scaffold for tissue engineering in oral surgery. Stem Cells Int. **2017**, 4585401 (2017)
5. Jaipaew, J., Wangkulangkul, P., Meesane, J., Raungrut, P., Puttawibul, P.: Mimicked cartilage scaffolds of silk fibroin/hyaluronic acid with stem cells for osteoarthritis surgery: morphological, mechanical, and physical clues. Mater Sci Eng C Mater Biol Appl. **64**, 173–182 (2016)

6. Cheema, S.K., Chen, E., Shea, L.D., Mathur, A.B.: Regulation and guidance of cell behavior for tissue regeneration via the siRNA mechanism. Wound Repair Regen. **15**(3), 286–295 (2007)
7. Howard, D., Buttery, L.D., Shakesheff, K.M., Roberts, S.J.: Tissue engineering: strategies, stem cells and scaffolds. J Anat. **213**(1), 66–72 (2008)
8. Freitas, G.P., Lopes, H.B., Souza, A.T.P., Oliveira, P.G.F.P., Almeida, A.L.G., Souza, L.E.B., Coelho, P.G., Beloti, M.M., Rosa, A.L.: Cell therapy: effect of locally injected mesenchymal stromal cells derived from bone marrow or adipose tissue on bone regeneration of rat calvarial defects. Sci Rep. **9**(1), 13476 (2019)
9. Brown, P.T., Handorf, A.M., Jeon, W.B., Li, W.J.: Stem cell-based tissue engineering approaches for musculoskeletal regeneration. Curr Pharm Des. **19**(19), 3429–3445 (2013)
10. Paschos, N.K., Brown, W.E., Eswaramoorthy, R., Hu, J.C., Athanasiou, K.A.: Advances in tissue engineering through stem cell-based co-culture. J. Tissue Eng. Regen. Med. **9**(5), 488–503 (2015)
11. Mustoe, T.A., Pierce, G.F., Morishima, C., Deuel, T.F.: Growth factor-induced acceleration of tissue repair through direct and inductive activities in a rabbit dermal ulcer model. J Clin Invest. **87**(2), 694–703 (1991)
12. Augustyniak, E., Trzeciak, T., Richter, M., Kaczmarczyk, J., Suchorska, W.: The role of growth factors in stem cell-directed chondrogenesis: a real hope for damaged cartilage regeneration. Inter Ortho. (SICOT) **39**, 995–1003 (2015)
13. Vo, T.N., Kasper, F.K., Mikos, A.G.: Strategies for controlled delivery of growth factors and cells for bone regeneration. Adv Drug Deliv Rev. **64**(12), 1292–1309 (2012)
14. Park, U., Kim, K.: Multiple growth factor delivery for skin tissue engineering applications. Biotechnol Bioproc E. **22**, 659–670 (2017)
15. Werner, S., Grose, R.: Regulation of wound healing by growth factors and cytokines. Physiol Rev. **83**, 835–870 (2003)
16. Yi, S., Ding, F., Gong, L., Gu, X.: Extracellular matrix scaffolds for tissue engineering and regenerative medicine. Curr Stem Cell Res. **12**(3), 233–246 (2017)
17. Collins, M.N., Ren, G., Young, K., Pina, S., Reis, R.L., Oliveira, J.M.: Scaffold fabrication technologies and structure/function properties in bone tissue engineering. Adv Funct Mater. **31**, 2010609 (2021)
18. Gómez, S., Vlad, M.D., López, J., Fernández, E.: Design and properties of 3D scaffolds for bone tissue engineering. Acta Biomater. **42**, 341–350 (2016)
19. Bitar, K.N., Zakhem, E.: Design strategies of biodegradable scaffolds for tissue regeneration. Biomed Eng Comput Biol. **6**, 13–20 (2014)
20. Liu, C.Z., Czernuszka, J.T.: Development of biodegradable scaffolds for tissue engineering: a perspective on emerging technology. Mater Sci Tech-Lond. **23**(4), 379–391 (2007)
21. Yan, F., Liu, Y., Chen, H., Zhang, F., Zheng, L., Hu, Q.: multi-scale controlled tissue engineering scaffold prepared by 3D printing and NFES technology. AIP Adv. **4**, 031321 (2014)
22. Benyus, J.M.: Biomimicry: innovation inspired by nature. Harper Perennial, New York (2002)
23. Naik, R.R., Singamaneni, S.: Introduction: bioinspired and biomimetic materials. Chem. Rev. **117**(20), 12581–12583 (2017)
24. Reddy, R., Reddy, N.: Biomimetic approaches for tissue engineering. J Biomat Sci-Poym E. **29**, 1667–1685 (2018)
25. Grayson, W.L., Martens, T.P., Eng, G.M., Radisic, M., Vunjak-Novakovic, G.: Biomimetic approach to tissue engineering. Semin Cell Dev Biol. **20**(6), 665–673 (2009)
26. Gao, Q., Xie, C., Wang, P., Xie, M., Li, H., Sun, A., Fu, J., He, Y.: 3D printed multi-scale scaffolds with ultrafine fibers for providing excellent biocompatibility. Mat Sci Eng C-Mater. **107**, 110269 (2020)
27. Freeman, F.E., Browe, D.C., Diaz-Payno, J., Nulty, J., Euw, S., Grayson, W.L., Kelly, D.J.: Biofabrication of multiscale bone extracellular matrix scaffolds for bone tissue engineering. Eur Cells Mater. **38**, 168–187 (2019)
28. Smith, L.A., Liu, X., Ma, P.X.: Tissue engineering with nano-fibrous scaffolds. Soft Matter **4**(11), 2144–2149 (2008)

29. Kyle, S., Aggeli, A., Ingham, E., McPherson, M.J.: Production of self-assembling biomaterials for tissue engineering. Trends Biotechnol. **27**, 423–433 (2009)
30. Laurent, C.P., Ganghoffer, J.F., Babin, J., Six, J.L., Wang, X., Rahouadj, R.: Morphological characterization of a novel scaffold for anterior cruciate ligament tissue engineering. J Biomech Eng. **133**(6), 065001 (2011)
31. Cheng, A., Schwartz, Z., Kahn, A., Li, X., Shao, Z., Sun, M., Ao, Y., Boyan, B.D., Chen, H.: Advances in porous scaffold design for bone and cartilage tissue engineering and regeneration. Tissue Eng Part B-Re. **25**, 14–29 (2019)
32. Ramakrishna, H., Li, T., He, T., Temple, J., King, M.W., Spagnoli, A.: Tissue engineering a tendon-bone junction with biodegradable braided scaffolds. Biomater Res. **23**, 11 (2019)
33. Panjapheree, K., Kamonmattayakul, S., Meesane, J.: Biphasic scaffolds of silk fibroin film affixed to silk fibroin/chitosan sponge based on surgical design for cartilage defect in osteoarthritis. Mater Design **141**, 323–332 (2018)
34. Webber, M.J., Khan, O.F., Sydlik, S.A., Tang, B.C., Langer, R.: A Perspective on the clinical translation of scaffolds for tissue engineering. Ann Biomed Eng **43**, 641–656 (2015)
35. Chan, B.P., Leong, K.W.: Scaffolding in tissue engineering: general approaches and tissue-specific considerations. Eur Spine J. **17**(Suppl 4), 467–479 (2008)
36. Tallawi, M., Rosellini, E., Barbani, N., Cascone, M.G., Rai, R., Saint-Pierre, G., Boccaccini, A.R.: Strategies for the chemical and biological functionalization of scaffolds for cardiac tissue engineering: a review. J R Soc Interface. **12**(108), 20150254 (2015)
37. Zhou, Z.X., Chen, Y.R., Zhang, J.Y., Jiang, D., Yuan, F.Z., Mao, Z.M., Yang, F., Jiang, W.B., Wang, X., Yu, J.K.: Facile strategy on hydrophilic modification of poly(ε-caprolactone) scaffolds for assisting tissue-engineered meniscus constructs in vitro. Front Pharmacol. **11**, 471 (2020)
38. Zhu, J., Marchant, R.E.: Design properties of hydrogel tissue-engineering scaffolds. Expert Rev Med Devices. **8**(5), 607–626 (2011)
39. Ghodbane, S.A., Dunn, M.G.: Physical and mechanical properties of cross-linked type I collagen scaffolds derived from bovine, porcine, and ovine tendons. J Biomed Mater Res A. **104**(11), 2685–2692 (2016)
40. Collins, M.N., Ren, G., Young, K., Pina, S., Reis, R.L., Oliveira, J.M.: Scaffold fabrication technologies and structure/function properties in bone tissue engineering. Adv Healthc Mater. **31**, 2010609 (2021)
41. Rao, S.S., Winter, J.O.: Adhesion molecule-modified biomaterials for neural tissue engineering. Front Neuroengineering. **2**, 6 (2009)
42. Zhu, J.: Bioactive modification of poly(ethylene glycol) hydrogels for tissue engineering. Biomaterials **31**, 4639–4656 (2010)
43. Tlatli, R., Nozach, H., Collet, G., Beau, F., Vera, L., Stura, E., Dive, V., Cuniasse, P.: Grafting of functional motifs onto protein scaffoldsidentified by PDB screening—an efficient route to designoptimizable protein binders. FEBS J. **280**, 139–159 (2013)
44. Rai, R., Raval, R., Khandeparker, R.V.S., Chidrawar, S.K., Khan, A.A., Ganpat, M.S.: Tissue engineering: step ahead in maxillofacial reconstruction. J. Int. Oral. Health. **7**(9), 138–142 (2015)
45. Rodella, L.F., Favero, G., Labanca, M.: Biomaterials in maxillofacial surgery: membranes and grafts. Int. J. Biomed. Sci. **7**(2), 81–88 (2011)
46. Lee, S., Goh, B.T., Tideman, H., Stoelinga, P.J.W.: Modular endoprosthesis for mandibular reconstruction: a preliminary animal study. Int. J. Oral. Maxillofac. Surg. **37**(10), 935–942 (2008)
47. Nguyen, T.T.H., Eo, M.Y., Myoung, H., Kim, M.J., Min Kim, S.M.: Implant-supported fixed and removable prostheses in the fibular mandible. Int. J. Implant. Dent. **6**(44) (2020).https://doi.org/10.1186/s40729-020-00241-7
48. Bitar, D., Parvizi, J.: Biological response to prosthetic debris. World J. Orthop. **6**(2), 172–189 (2015)
49. Coen, N., Kadhim, M.A., Wright, E.G., Case, C.P., Mothersill, C.E.: Particulate debris from a titanium metal prosthesis induces genomic instability in primary human fibroblast cells. British J. Cancer **88**, 548–552 (2003)

50. Wong, R.C.W., Kin, T.L., Merkx, M.A.W.: Biomechanics of mandibular reconstruction: a review. Int. J Oral Max Surg. **39**, 313–319 (2010). Biodegradable Materials and Metallic Implants—A Review
51. Prakasam, M., Locs, J., Salma-Ancane, K., Loca, D., Largeteau, A., Berzina-Cimdina, L.: J Funct Biomater. **8**(4), 44 (2017)
52. Sándor, G.K.B.: Tissue engineering of bone: Clinical observations with adipose-derived stem cells, resorbable scaffolds, and growth factors. Ann Maxillofac Surg. **2**(1), 8–11 (2012)
53. Pacifici, A., Polimeni, A., Pacifici, L.: Additive manufacturing and biomimetic materials in oral and maxillofacial surgery: a topical overview. J Biol Regul Homeost Agents. **32**(6), 1579–1582 (2018)
54. Cordonnier, T., Sohier, J., Rosset, P., Layrolle, P.: Biomimetic materials for bone tissue engineering—State of the art and future trends. Adv Eng Mater. **13**, B135–B150 (2011)
55. Muallah, D., Sembdner, P., Holtzhausen, S., Meissner, H., Hutsky, A., Ellmann, D., Assmann, A., Schulz, M.C., Lauer, G., Kroschwald, L.M.: Adapting the pore size of individual, 3D-printed CPC scaffolds in maxillofacial surgery. J. Clin. Med. **10**(12), 2654 (2021)
56. Hosseinpour, S., Ahsaie, M.G., Rad, M.R., Baghani, M., Motamedian, S.R., Khojasteh, A.: Application of selected scaffolds for bone tissue engineering: a systematic review. Oral Maxillofac Surg. **21**, 109–129 (2017)
57. Sangkert, S., Kamolmatyakul, S., Gelinsky, M., Meesane, J.: 3D printed scaffolds of alginate/polyvinylalcohol with silk fibroin based on mimicked extracellular matrix for bone tissue engineering in maxillofacial surgery **26**, 102140 (2021)
58. Miszuk, J.M., Hu, J., Sun, H.: Biomimetic nanofibrous 3D materials for craniofacial bone tissue engineering. ACS Appl. Bio Mater. **3**(10), 6538–6545 (2020)
59. Liua, X., Smith, L.A., Hua, J., Ma, P.X.: Biomimetic nanofibrous gelatin/apatite composite scaffolds for bone tissue engineering. Biomaterials **30**, 2252–2258 (2009)
60. Gentile, P., Mattioli-Belmonte, M., Chiono, V., Ferretti, C., Baino, F., Tonda-Turo, C., Vitale-Brovarone, C., Pashkuleva, I., Reis, R.L., Ciardelli, G.: Bioactive glass/polymer composite scaffolds mimicking bone tissue. J Biomed Mater Res A. **100A**, 2654–2667 (2012)
61. Frohbergh, M.E., Katsman, A., Botta, G.P., Lazarovici, P., Schauer, C.L., Wegst, U.G.K., Lelkes, P.I.: Electrospun hydroxyapatite-containing chitosan nanofibers crosslinked with genipin for bone tissue engineering. Biomaterials **33**, 9167–9178 (2012)
62. Lee, H., Lim, S., Birajdar, M.S., Lee, S.H., Park, H.: Fabrication of FGF-2 immobilized electrospun gelatin nanofibers for tissue engineering. Int J Biol Macromol. **93**, 1559–1566 (2016)
63. Malafay, P.B., Silva, G.A., Reis, R.L.: Natural–origin polymers as carriers and scaffolds for biomolecules and cell delivery in tissue engineering applications. Adv Drug Deliver Rev. **59**, 207–233 (2007)
64. Guo, J.L., Kim, Y.S., Mikos, A.G.: Biomacromolecules for tissue engineering: emerging biomimetic strategies. Biomacromol **20**(8), 2904–2912 (2019)
65. Grande, D., Ramier, J., Versace, D.L., Renard, E., Langlois, V.: Design of functionalized biodegradable PHA-based electrospun scaffolds meant for tissue engineering applications. New Biotechnol. **37**, 129–137 (2017)
66. Cole, B.J., Pascual-Garrido, C., Grumet, R.C.: Surgical management of articular cartilage defects in the knee. Instr Course Lect. **59**, 181–204 (2010)
67. Kwon, H., Brown, W.E., Lee, C.A., Wang, D., Paschos, N., Hu, J.C., Athanasiou, K.A.: Surgical and tissue engineering strategies for articular cartilage and meniscus repair. Nat Rev Rheumatol. **15**(9), 550–570 (2019)
68. Kellett, C.F., Boscainos, P.J., Gross, A.E.: Surgical options for articular defects of the knee. Expert Rev Med Devices. **3**(5), 585–593 (2006)
69. Mihalko, W.M., Haider, H., Kurtz, S., Marcolongo, M., Urish, K.: New materials for hip and knee joint replacement: What's hip and what's in kneed? J Ortho Res. **37**, 1436–1444 (2020)
70. Sloten, J.V., Labey, L., Van Audekercke, R., Van der Perre, G.: Materials selection and design for orthopaedic implants with improved long-term performance. Biomaterials **19**(16), 1455–1459 (1998)

71. Frenkel, S.R., Di Cesare, P.E.: Scaffolds for articular cartilage repair. Ann Biomed Eng. **32**, 26–34 (2004)
72. Zhao, Z., Fan, C., Chen, F., Sun, Y., Xia, Y., Ji, A., Wang, D.A.: Progress in articular cartilage tissue engineering: a review on therapeutic cells and macromolecular scaffolds. Macromol Biosci. **20**, 1900278 (2019)
73. Jaipaew, J., Wangkulangkul, P.: Meesane J*, Raungrut P, Puttawibul P, Mimicked cartilage scaffolds of silk fibroin/hyaluronic acid with stem cells for osteoarthritis surgery: morphological, mechanical, and physical clues. Mat Sci Eng C-Mater. **64**, 173–182 (2016)
74. Wasyłeczko, M., Sikorska, W., Chwojnowski, A.: Review of synthetic and hybrid scaffolds in cartilage tissue engineering. Membranes **10**, 380 (2020)
75. Bell, E.: Strategy for the selection of scaffolds for tissue engineering. Tissue Eng. **1**(2), 163–79 (1995)
76. Koh, Y.G., Lee, J.A., Kim, Y.S., Lee, H.Y., Kim, H.J., Kan, K.T.: Optimal mechanical properties of a scaffold for cartilage regeneration using finite element analysis. J Tissue Eng. **10**, 2041731419832133 (2019)
77. Smeriglio, P., Lai, J.H., Yang, F., Bhutani, N.: 3D Hydrogel scaffolds for articular chondrocyte culture and cartilage generation. J Vis Exp. **104**, 53085 (2015)
78. Wei, W., Ma, Y., Yao, X., Zhou, W., Wang, X., Li, C., Lin, J., He, Q., Leptihn, S., Ouyang, H.: Advanced hydrogels for the repair of cartilage defects and regeneration. Bioact Mater. **6**(4), 998–1011 (2021)
79. Fu, Y., Zoetebier, B., Both, S., Dijkstra, P.J., Karperien, M.: Engineering of optimized hydrogel formulations for cartilage repair. Polymers **13**, 1526 (2021)
80. Chen, G., Kawazoe, N.: Porous scaffolds for regeneration of cartilage, bone and osteochondral tissue. Adv. Exp. Med. Biol. **1058**, 171–191 (2018)

# Chapter 2
# Principles of Tissue Engineering

Tissue engineering is an attractive approach for biomedical technology. At the beginning of tissue engineering, around the 1980s, biodegradable polymers were fabricated into a "scaffold", before being cultured with cells that regulated themselves into the proposed tissue. Then, the cultured scaffold was inserted to the body, to replace the tissue defect [1, 2]. From this development, tissue engineering is currently a significant issue for biomedical application. Principally, tissue engineering is applied in three main areas: (1) science, (2) medicine, and (3) engineering. The basic principles of tissue engineering are shown in Fig. 2.1. The important components to construct tissue are classified into three parts: (1) cells, (2) growth factors, and (3) scaffolds [3, 4].

## 2.1 Cells and Their Function

Cells play the important role of self-regulating into the proposed tissue. Although cells are only small units of the body, they serve many important roles and are involved with many mechanisms of the human body. Irregular function of cells is a cause of many diseases, which in turn lead to the dysfunction of tissue and organs [5]. Generally, human cells are composed of three main parts: (1) the nucleus, (2) cytoplasm, and (3) the cell membrane, and within these, each part has many smaller components (Fig. 2.2).

Generally, for tissue engineering, stem cells and primary cells are used to construct tissue. Stem cells were divided into two types: (1) pluripotent stem or embryonic stem cells and (2) multipotent stem cells or adult stem cells [6] (Fig. 2.3). First, pluripotent stem cells are generated in the blastocyst period, during embryonic development (Fig. 2.3a). These pluripotent stem cells have a high potential to regulate into many types of primary cells. With this great potential, pluripotent stem cells can be used to repair defective tissue, via direct implantation at the target site. On the other



J. Meesane, *Mimicked Tissue Engineering Scaffolds for Maxillofacial and Articular Cartilage Surgery*, Engineering Materials,
https://doi.org/10.1007/978-981-19-7830-2_2

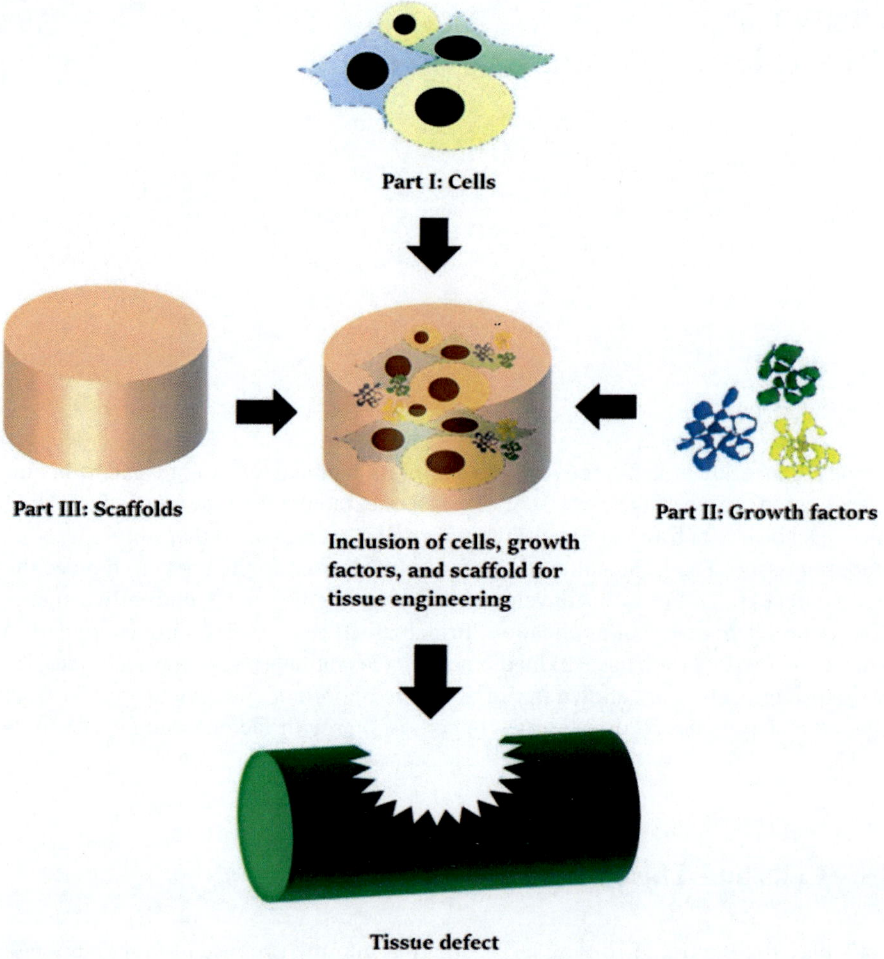


**Fig. 2.1** Concept of tissue engineering

hand, those pluripotent stem cells can be cultured on performance scaffolds that contain clues to induce tissue regeneration. Pluripotent stem cells can also regulate themselves into tumor cells. Therefore, irregular regulation has to be intensively considered before pluripotent stem cells are used for tissue engineering.

Second, multipotent stem cells are generated from organs (Fig. 2.3b). Generally, multipotent stem cells are in many parts of tissue organs, for instance, bone marrow, blood, the skin, cord blood, and the umbilical cord. However, multipotent stem cells have the limitation to regulate into primary cells.

Pluripotent and multipotent stem cells can function to self-regulate into primary cells, so as to repair defective tissue. Unfortunately, in the cases of the human body, there are insufficient multipotent stem cells to repair defective tissue. Therefore, in

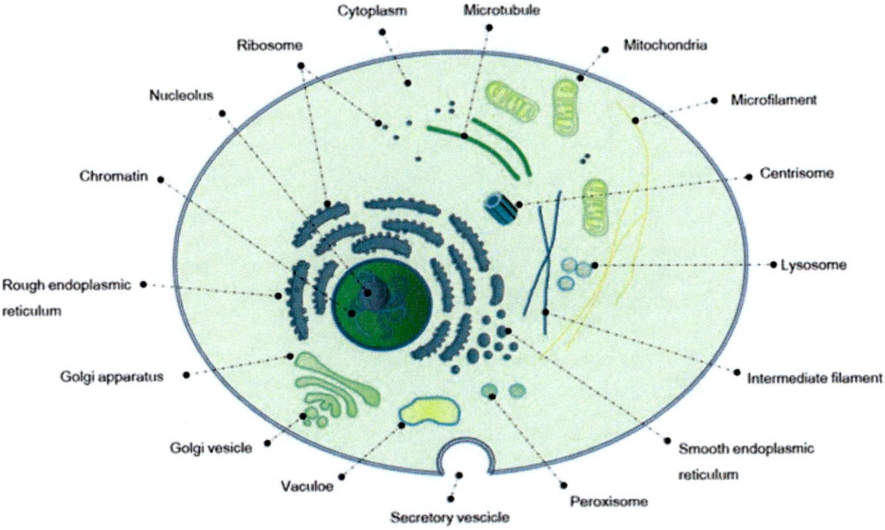


**Fig. 2.2** Anatomy of a human cell

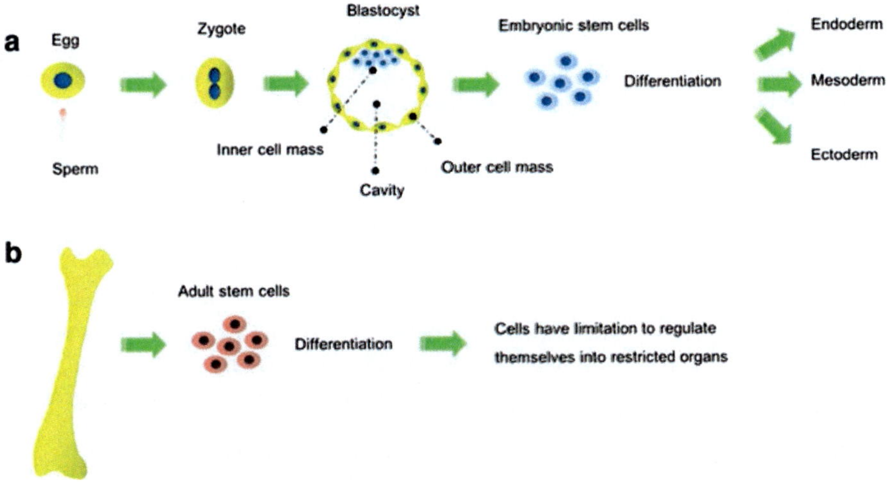


**Fig. 2.3** Classification of stem cells: **a** embryonic stem cells or pluripotent stem cells and **b** adult stem cells or multipotent stem cells

the areas of a large defect, tissue regeneration is not complete, which leads to scar formation.

In tissue engineering, the number of stem cells needs to be expanded, to enhance the potential to induce tissue regeneration at the defect site (Fig. 2.4). Besides expanding the number of stem cells, biological signals are combined with the stem cells to enhance tissue regeneration. Therefore, understanding the principles of

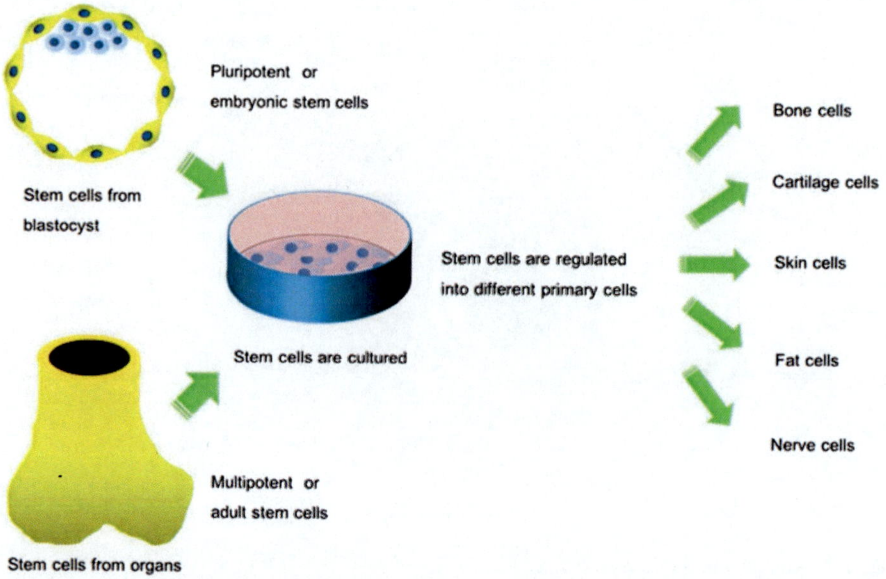


**Fig. 2.4** Expansion of stem cells before application in tissue engineering

growth factors for tissue engineering is important, so as to create high performance systems to regenerate new tissue at the defect site.

## 2.2 Growth Factors and Their Function

Growth factors are biological molecules that induce tissue regeneration. Some scientists call these growth factors as "soluble signals". In the molecules of growth factors, amino sequences, such as RGD and YIGSR, can be recognized by the cells; whereas, in tissue regeneration, there are many growth factors (Table 2.1), and each growth factor performs differently to induce tissue regeneration.

Generally, receptors of cells can attach to growth factors. The cells then respond to the growth factors, by organizing the components in their cytoplasm [17]. During this process, the molecules in the cytoplasm organize themselves as a pathway to connect between the growth factors and the nucleus. The molecules act as signals that are transferred into the nucleus. The responses of the nucleus to these signals are as follows: cell proliferation, migration, regulation, and morphogenesis [18, 19]. In other cases, the nucleus responds to these signals by producing some RNA molecules, which act as codes to produce proteins. RNA molecules are transferred to the ribosomes in the cytoplasm. Ribosomes have the important role of synthesizing proteins by decoding from the RNA molecules. Some proteins synthesized from cells become extracellular matrix components. The extracellular matrix performs

**Table 2.1** Examples of growth factors and their function

| Growth factors | Function |
|---|---|
| Bone morphogenetic protein (BMP) | To involve the skeletal development [7] |
| Epidermal growth factor (EGF) | To stimulate the growth of epidermal and epithelial cells [8] |
| Fibroblast growth factor (FGF) | To involve in proliferation, migration, and differentiation of cell [9] |
| Hepatocyte growth factor (HGF) | To regulate growth, motility, and morphogenesis of cell [10] |
| Insulin-like growth factor (IGF) | To involve in the systemic of body growth [11, 12] |
| Nerve growth factor (NGF) | To regulate the growth of nerve cells [13] |
| Platelet-derived growth factor (PDGF) | To promote wound healing and growth of stem cells [14] |
| Transforming growth factor (TGF) | To stimulate and inhibit the cell growth [15] |
| Vascular endothelial growth factor (VEGF) | To stimulate the formation of blood vessel [16] |

as an insoluble signal to induce tissue regeneration [20]. Some reports indicated that growth factors (soluble signals) coupled with the extracellular matrix (insoluble signals) work together in a harmonious, biological function to induce tissue regeneration [21].

## 2.3 Extracellular Matrix and Its Function

The extracellular matrix acts as the structure for cell adhesion in tissue [22], and it can also enhance mechanical stability of the tissue [23]. Furthermore, the extracellular matrix performs as an insoluble signal that can induce various cell behaviors in tissue, for instance: cell adhesion, proliferation, migration, regulation as well as morphogenesis, and these cell behaviors can in turn have the effect of inducing tissue regeneration.

The extracellular matrix is classified principally into three parts by its function in tissue: (1) reinforcement, (2) texture, and (3) attachment. First, reinforcement enhances the mechanical properties and maintains stability of the tissue [24]. The main components for reinforcement are collagen and elastin. Second, texture is the matrix of the tissue. Texture can maintain the shape of the tissue and resist compressive forces. The major components of texture are the glycosaminoglycans (for instance, hyaluronic acid, chondroitin sulfate, heparan sulfate, and keratan sulfate). The third part is attachment, which is the bridge to connect the components. Attachment can involve construction of the extracellular matrix into a sophisticated network. The network structure and all components of the extracellular matrix are shown in Fig. 2.5 and in Table 2.2.

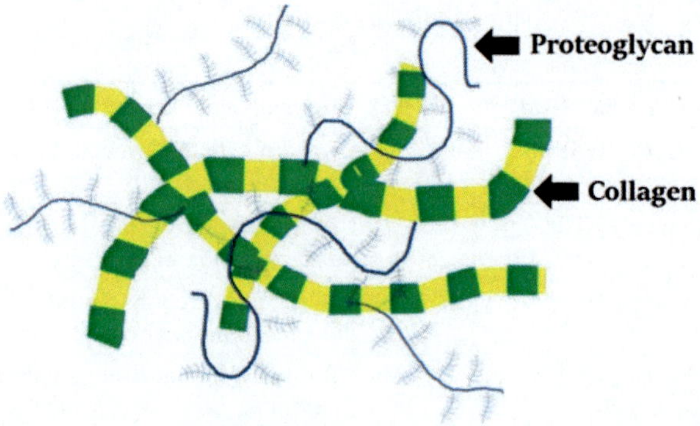


**Fig. 2.5** Structure of the extracellular matrix

**Table 2.2** Examples of extracellular matrix components and function

| Matrix | Function |
|---|---|
| Betaglycan | Receptor of TGFβ type III [25] |
| Chondroitin sulfate | Supporting the strength of cartilage, tendon, ligament, and aorta [26] |
| Collagen Type I,II,III,V | Main structural protein to support resident of cells [27] |
| Dermatan sulfate | Involvement in the role of coagulation, wound repairing [28] |
| Decorin | Growth factor binding [29] |
| Elastin | Function on elasticity of tissue [30] |
| Fibronectin | Binding with protein, adhesion, growth, migration, and differentiation of cells [31] |
| Heparin | Involvement in anticoagulation [32] |
| Heparan sulfate | Binding with proteins, involved in the development process, blood coagulation, and angiogenesis [33] |
| Hyaluronic acid | Supporting the lubrication of the joints [34] |
| Keratan sulfate | Supporting for absorption of the force in cornea, cartilage, and bone [35] |
| Laminin | Network structure in basal lamina, which resist tensile force [36] |
| Syndecan | Binding of growth factors, enhancement of activity [37] |
| Tenascin | Inducing of wound healing and cell adhesion controlling [38] |
| Versican | Binding of collagen and hyaluronic acid (HA) [39] |

## 2.4 Application of Cells in Tissue Engineering

Many types of cells have been cultured for applications in tissue engineering, before transplantation into site defects. To engineer a tissue, cells are seeded into scaffolds that have the functionality to support cell adhesion, proliferation, and migration, which in turn lead to tissue regeneration [40]. Generally, primary cells or stem cells are used for tissue engineering. To engineer the tissue, as a native structure, cells have been co-cultured into two or three dimensions, whichever is suitable for a given type of tissue. For instance, skin tissue has two major layers of cells: fibroblasts and keratinocytes. To construct this structure, fibroblasts and keratinocytes are cultured and connected via the membrane scaffolds that function as an extracellular matrix (Fig. 2.6a). In the case of three-dimensional structures as bone tissue, osteoblasts and osteoclasts are seeded into three-dimensional scaffolds. In the practical approach, osteoblasts and osteoclasts, with or without encapsulation, are directly seeded and cultured in the scaffolds (Fig. 2.6b). For encapsulation, cells are protected from "non-pleasure circumstances" during culturing, particularly in the complicated system of culturing in bioreactors (Fig. 2.6c). Those circumstances have an effect on the functional expression of the cells that lead to a low potential for tissue regeneration. Therefore, encapsulating the cells is a suitable approach to enhance the performance of tissue regeneration.

Besides constructing tissue, as a native structure, co-culturing has been used to evaluate cell regulation in tissue diseases that lead to diagnostics as well as screening for molecular or gene therapy [41]. For instance, co-culturing was created to be used as models to evaluate these diseases [42, 43]. To create these models, disease cells are co-cultured with normal cells. Then, the secreted signals during culturing of those cells are analyzed to evaluate progression of cancer [44]. The model of co-culturing is used as a guide to search for suitable bioactive molecules to suppress said disease expression [45].

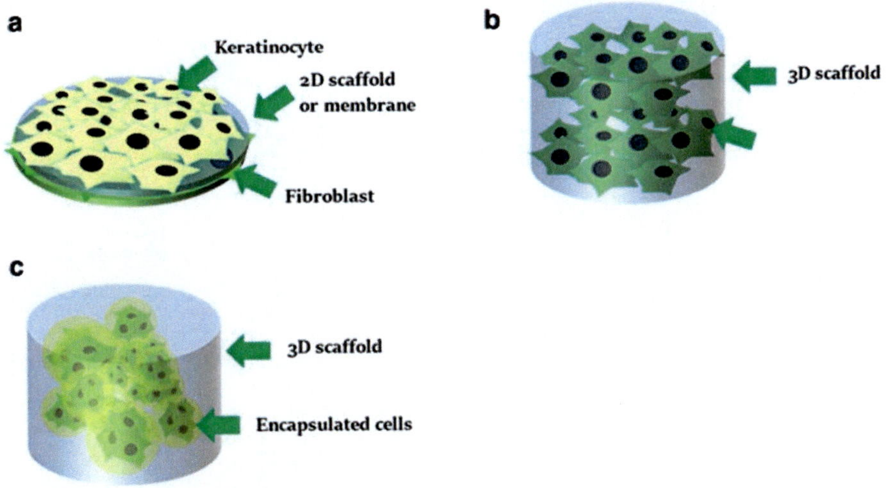


**Fig. 2.6** **a** Construction of tissue into a two-dimensional structure, **b** construction of tissue into a three-dimensional structure, without **c**, and with cell encapsulation

## 2.5 Application of Growth Factors in Tissue Engineering

Growth factors have been often added into the culturing system in the media solution in tissue engineering. These factors, without the carriers, when combined directly with the cultured cells have an effect on cell regulation in the proposed tissue [46]. On the other hand, when the growth factors are added into the delivery system by carriers [47], the carriers are combined by physical or chemical interaction, which could release said factors at the target sites [48, 49]. The enhancement of specific groups in the molecular structure of carriers is an attractive approach, for enhancing the potential of cell recognition at the target sites [50]. Therefore, it is important in tissue regeneration to create a high performance carrier system for growth factor delivery.

In the system of tissue regeneration, there are many growth factors that synergize to induce tissue formation. To create a releasing system as native tissue regeneration is an attractive approach for growth factor applications. In different systems of tissue regeneration, there are different sequences of growth factor release. To create the sequences of growth factor release, as in native circumstances, can promote tissue regeneration [51]. Designing the sequences for growth factors, that are activated in the early stage of tissue regeneration, is connected with carriers with weaker interactions than in the later stages of tissue regeneration. The designed sequences of growth factors are shown in Fig. 2.7.

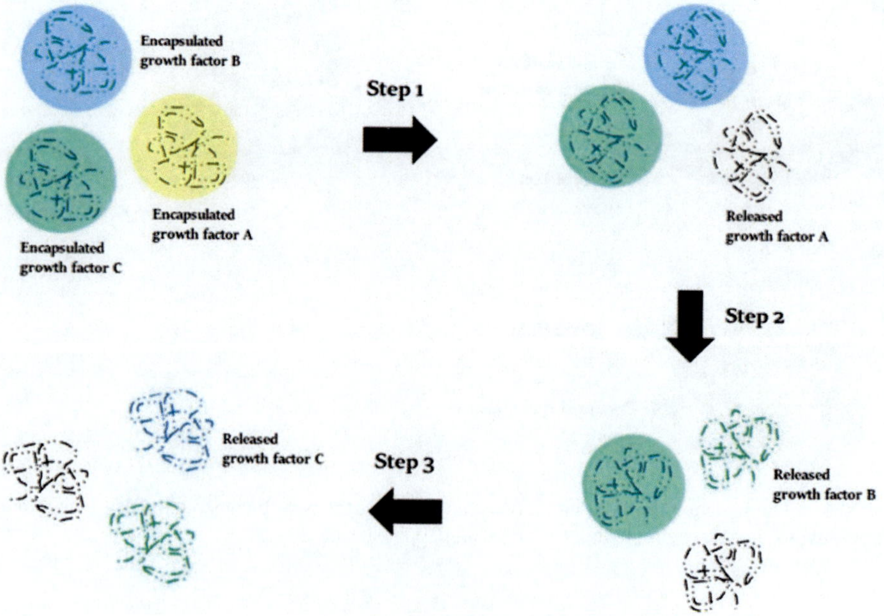


**Fig. 2.7** Designed sequences of growth factors for tissue regeneration

Besides growth factors that are used for tissue engineering, some can be used in combination with other biological molecules, or some genes for the purpose of enhancing tissue regeneration. Furthermore, growth factors can be used with some biological molecules as therapy agents for some diseases, which, in turn, leads to creating a therapy system for diseases. The significant point to be emphasized is to have an understanding of the functionality of growth factors and biological molecules, before designing or creating any system.

## 2.6 Application of Extracellular Matrix in Tissue Engineering

The extracellular matrix is an important part of tissue that performs as a native scaffold [52]. Generally, decellularized tissue has been used for native scaffolds in cell culturing [53]. Decellularized tissue has the biological signals to induce cell regulation in the proposed tissue. However, the critical problem that often exists in decellularized tissue is the immune rejection in in vivo testing [54]. Therefore, to totally remove whole cells from the tissue, without the residual molecules that cause immune rejection, is important for application in tissue engineering [55].

Furthermore, the structure and function of decellularized tissue have been destroyed which leads to a fragment formation. Therefore, in some cases, biological molecules have been added in decellularized tissue, so as to reconstruct their structure and functionality [ref] (Fig. 2.8a). For instance; fibronectin has been added to combine with those fragments of decellularized tissue [56]. In some cases, collagen has been combined with decellularized tissue to reconstruct the reinforcement of the extracellular matrix [57].

Besides constructing the extracellular matrix to be similar to the native structure by using decellularized tissue, construction by mixing with trigger molecules is an alternative choice [58]. Collagen has been used often as the base material for construction [59] (Fig. 2.8b). For this, the adjusting condition for self-organization of collagen in the fibril structure similar to the extracellular matrix is emphasized. This can be used as scaffolds for tissue regeneration. Furthermore, self-organization of collagen can combine with other trigger molecules that can enhance performance similar to the extracellular matrix, for instance, fibronectin, laminine, and glycosaminoglycan. These trigger molecules have important roles for the induction of tissue regeneration [60].

Another choice, that has been used to enhance the performance of these constructed scaffolds, is adding polymers. Polymers (e.g., chitosan, alginate, cellulose derivative, polycarpolectone, and polylactic acid) have often shown functionality to enhance tissue regeneration [61]. Generally, those polymers can induce cell adhesion, proliferation, and biodegradation. The polymers are also able to form hydrogel for inducing of swelling behavior [62].

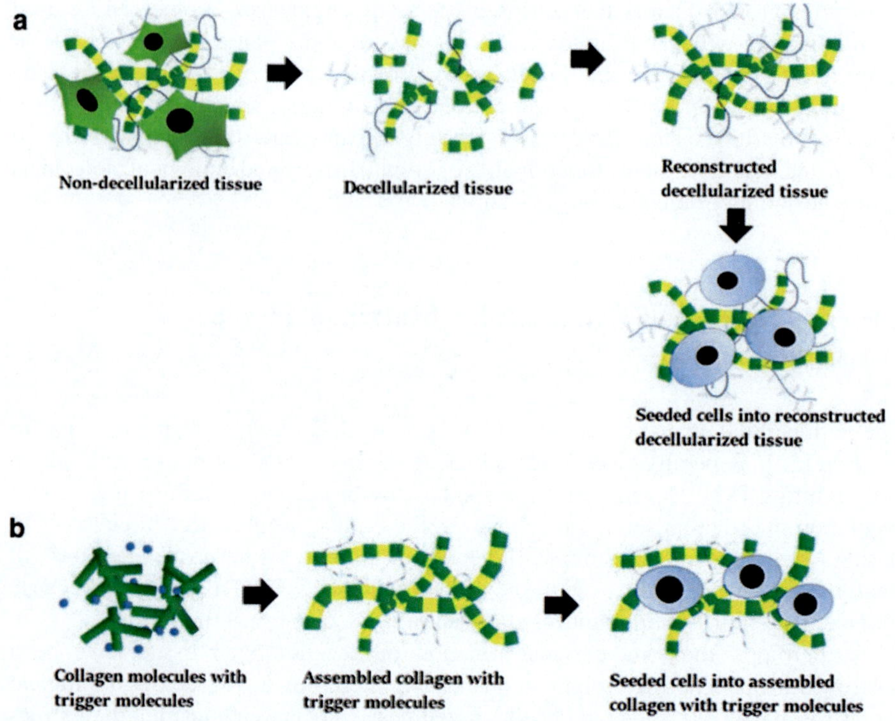


**Fig. 2.8** Construction of scaffolds: **a** construction based on reconstruction of decellularized tissue and **b** construction based on assembly of collagen

According to the contents in this chapter, the principles of tissue engineering regarding their three parts: cells, growth factors, and extracellular matrix were explained. The instances of design in tissue engineering were shown as the guidance in the creation of performance tissue engineering. To promote the performance, mimicking of tissue engineering will be explained in the next chapter.

## References

1. Patrick, C.W., Jr., Zheng, B., Johnston, C., Reece, G.P.: Long-term implantation of preadipocyte-seeded PLGA scaffolds. Tissue Eng. Pt A **8**, 283–293 (2002)
2. Hirsch, H., Laemmle, C., Behr, B., Lehnhardt, M., Jacobsen, F., Hoefer, D., Maximilian, K.: Implant for autologous soft tissue reconstruction using an adipose-derived stem cell-colonized alginate scaffold. J. Plast. Reconstr. Aesthet. Surg. **71**(1), 101–111 (2018)
3. Tollemar, V., Collier, Z.J., Mohammed, M.K., Lee, M.J., Ameer, G.A., Reid, R.R.: Stem cells, growth factors and scaffolds in craniofacial regenerative medicine. Genes Dis. **3**(1), 56–71 (2016)
4. Shokeir, A.A., Harraz, A.M., Shehab El-Din, A.B.: Tissue engineering and stem cells: basic principles and applications in urology. Int. J. Urol. **17**, 964–973 (2010)

5. Egeblad, M., Nakasone, E.S., Werb, Z.: Tumors as organs: complex tissues that interface with the entire organism. Dev. Cell **18**(6), 884–901 (2010)
6. Kwon, S.G., Kwon, Y.W., Lee, T.W., Park, G.T., Kim, J.H.: Recent advances in stem cell therapeutics and tissue engineering strategies. Biomater. Res. **22**, 36 (2018)
7. Katagiri, T., Watabe, T.: Bone morphogenetic proteins. Cold Spring Harb. Perspect. Biol. **8**(6), a021899 (2016)
8. Shu, D.Y., Hutcheon, A.E.K., Zieske, J.D., Guo, X.: Epidermal growth factor stimulates transforming growth factor-beta receptor Type II expression in corneal epithelial cells. Sci. Rep. **9**, 8079 (2019)
9. Yun, Y.R., Won, J.E., Jeon, E., Lee, S., Kang, W., Jo, H., Jang, J.H., Shin, U.S., Kim, H.W.: Fibroblast growth factors: biology, function, and application for tissue regeneration. J. Tissue Eng. **2010**, 218142 (2010)
10. Nakamura, T., Mizuno, S.: The discovery of Hepatocyte Growth Factor (HGF) and its significance for cell biology, life sciences and clinical medicine. Proc. Jpn Acad. Ser. B Phys. Biol. Sci. **86**(6), 588–610 (2010)
11. Lewitt, M.S., Boyd, G.W.: The role of insulin-like growth factors and insulin-like growth factor-binding proteins in the nervous system. Biochem. Insights **12**, 1178626419842176 (2019)
12. Talia, C., Connolly, L., Fowler, P.A.: The insulin-like growth factor system: A target for endocrine disruptors? Environ Int **147**, 106311 (2021)
13. Aloe, L., Rocco, M.L., Balzamino, B.O., Micera, A.: Nerve growth factor: a focus on neuroscience and therapy. Curr. Neuropharmacol. **13**(3), 294–303 (2015)
14. Wu, L.W., Chen, W.L., Huang, S.M., Chan, J.Y.H.: Platelet-derived growth factor-AA is a substantial factor in the ability of adipose-derived stem cells and endothelial progenitor cells to enhance wound healing. FASEB J. **33**, 2388–2395 (2019)
15. Moses, H.L., Yang, E.Y., Pietenpol, J.A.: TGF-beta stimulation and inhibition of cell proliferation: new mechanistic insights. Cell **63**(2), 245–247 (1990)
16. Bautch, V.L.: VEGF-directed blood vessel patterning: from cells to organism. Cold Spring Harb. Perspect. Med. **2**(9), a006452 (2012)
17. Jones, S., Kazlauskas, A.: Connecting signaling and cell cycle progression in growth factor-stimulated cells. Oncogene **19**, 5558–5567 (2000)
18. Matamales, M., Girault, J.A.: Signaling from the cytoplasm to the nucleus in striatal medium-sized spiny neurons. Front Neuroanat. **5**, 1–13 (2011)
19. Frieden, B.R., Gatenby, R.A.: Signal transmission through elements of the cytoskeleton form an optimized information network in eukaryotic cells. Sci. Rep. **9**, 6110 (2019)
20. Clause, K.C., Barker, T.H.: Extracellular matrix signaling in morphogenesis and repair. Curr. Opin. Biotechnol. **24**(5), 830–833 (2013)
21. Briquez, P.S., Hubbell, J.A., Martino, M.M.: Extracellular matrix-inspired growth factor delivery systems for skin wound healing. Adv. Wound Care (New Rochelle) **4**(8), 479–489 (2015)
22. Laudani, S., La Cognata, V., Iemmolo, R., Bonaventura, G., Villaggio, G., Saccone, S., Barcellona, M.L., Cavallaro, S., Sinatra, F.: Effect of a bone marrow-derived extracellular matrix on cell adhesion and neural induction of dental pulp stem cells. Front. Cell Dev. Biol. **8**, 100 (2020)
23. Ng, S.P., Billings, K.S., Ohashi, T., Allen, M.D., Best, R.B., Randles, L.G., Erickson, H.P., Clarke, J.: Designing an extracellular matrix protein with enhanced mechanical stability. PNAS **104**(23), 9633–9637 (2007)
24. Kular, J.K., Basu, S., Sharma, R.I.: The extracellular matrix: structure, composition, age-related differences, tools for analysis and applications for tissue engineering. J. Tissue Eng. **5**, 1–17 (2014)
25. Blair, C.R., Stone, J.B., Wells, R.G.: The type III TGF-β receptor betaglycan transmembrane-cytoplasmic domain fragment is stable after ectodomain cleavage and is a substrate of the intramembrane protease γ-secretase. Biochim. Biophys. Acta **1813**(2), 332–339 (2011)
26. Henrotin, Y., Mathy, M., Sanchez, C., Lambert, C.: Chondroitin sulfate in the treatment of Osteoarthritis: from in vitro studies to clinical recommendations. Ther. Adv. Musculoskelet Dis. **2**(6), 335–348 (2010)

27. Maurer, T., Stoffel, M.H., Belyaev, Y., Stiefel, N.G., Vidondo, B., Kueker, S., Mogel, H., Schaefer, B., Balmer, J.: Structural characterization of four different naturally occurring porcine collagen membranes suitable for medical applications. PLoS ONE **3**, 1–17 (2018)
28. Trowbridge, J.M., Gallo, R.L.: Dermatan sulfate: new functions from an old glycosaminoglycan. Glycobiology **12**(9), 117R-R125 (2002)
29. Santra, M., Reed, C.C., Iozzo, R.V.: Decorin binds to a narrow region of the epidermal growth factor (EGF) receptor, partially overlapping but distinct from the EGF-binding epitope. J. Biol. Chem. **277**(38), 35671–35681 (2002)
30. Green, E.M., Mansfield, J.C., Bell, J.S., Winlove, C.P.: The structure and micromechanics of elastic tissue. Interface Focus. **4**(2), 20130058 (2014)
31. Parisi, L., Toffoli, A., Ghezzi, B., Mozzoni, B., Lumetti, S., Macaluso, G.M.: A glance on the role of fibronectin in controlling cell response at biomaterial interface. Jpn. Dent. Sci. Rev. **56**(1), 50–55 (2020)
32. Oduah, E.I., Linhardt, R.J., Sharfstein, S.T.: Heparin: past, present, and future. Pharmaceuticals **9**(3), 38 (2016)
33. Chiodelli, P., Bugatti, A., Urbinati, C., Rusnati, M.: Heparin/heparan sulfate proteoglycans glycomic interactome in angiogenesis: biological implications and therapeutical use. Molecules **20**(4), 6342–6388 (2015)
34. Lin, W., Liu, Z., Kampf, N., Klein, J.: The role of hyaluronic acid in cartilage boundary lubrication. Cells **9**(7), 1606 (2020)
35. Caterson, B., Melrose, J.: Keratan sulfate, a complex glycosaminoglycan with unique functional capability. Glycobiology **28**(4), 182–206 (2018)
36. Han, R., Kanagawa, M., Yoshida-Moriguchi, T., Rader, E.P., Ng, R.A., Michele, D.E., Muirhead, D.E., Kunz, S., Moore, S.A., Iannaccone, S.T., Miyake, K., McNeil, P.L., Mayer, U., Oldstone, M.B.A., Faulkner, J.A., Campbell, K.P.: Basal lamina strengthens cell membrane integrity via the laminin G domain-binding motif of α-dystroglycan. PNAS **106**(31), 12573–12579 (2009)
37. Growth factors with enhanced syndecan binding generate tonic signalling and promote tissue healing. Nat. Biomed. Eng. **4**(4), 463–475 (2020)
38. Song, M.J., Hara, M., Imanaka-Yoshida, K., Lee, D.H., Chung, J.H., Lee, S.T.: Effects of tenascin C on the integrity of extracellular matrix and skin aging. Int. J. Mol. Sci. **21**, 8693 (2020)
39. Evanko, S.P., Potter-Perigo, S., Bollyky, P.L., Nepom, G.T., Wight, T.N.: Hyaluronan and versican in the control of human T-lymphocyte adhesion and migration. Matrix. Biol. **31**(2), 90–100 (2012)
40. Bueno, E.M., Laevsky, G., Barabino, G.A.: Enhancing cell seeding of scaffolds in tissue engineering through manipulation of hydrodynamic parameters. J. Biotechnol. **129**(3), 516–531 (2007)
41. Mountcastle, S.E., Cox, S.C., Sammons, R.L., Jabbari, S., Shelton, R.M., Kuehne, S.A.: A review of co-culture models to study the oral microenvironment and disease. J. Oral Microbiol. **12**(1), 1773122 (2020)
42. Kapałczyńska, M., Kolenda, T., Przybyła, W., Zajączkowska, M., Teresiak, A., Filas, V., Ibbs, M., Bliźniak, R., Łuczewski, L., Lamperska, K.: 2D and 3D cell cultures—a comparison of different types of cancer cell cultures. Arch. Med. Sci. **14**(4), 910–919 (2018)
43. Fujii, M., Sato, T.: Somatic cell-derived organoids as prototypes of human epithelial tissues and diseases. Nat. Mater. **20**, 156–169 (2021)
44. Morrison, C., Mancini, S., Cipollone, J., Kappelhoff, R., Roskelley, C., Overall, C.: Microarray and proteomic analysis of breast cancer cell and osteoblast co-cultures: role of osteoblast matrix metalloproteinase (MMP)-13 in bone metastasis. J. Biol. Chem. **286**(39), 34271–34285 (2011)
45. Edmondson, R., Broglie, J.J., Adcock, A.F., Liju, Y.L.: Three-dimensional cell culture systems and their applications in drug discovery and cell-based biosensors. Assay Drug Dev. Technol. **12**(4), 207–218 (2014)
46. Gross, S.M., Rotwein, P.: Unraveling growth factor signaling and cell cycle progression in individual fibroblasts. J Biol Chem. **291**(28), 14628–14638 (2016)

47. Atienza-Roca, P., Cui, X., Hooper, G.J., Woodfield, T.B.F., Lim, K.S.: Growth factor delivery systems for tissue engineering and regenerative medicine. Adv. Exp. Med. Biol. **1078**, 245–269 (2018)
48. Vo, T.N., Kasper, F.K., Mikos, A.G.: Strategies for controlled delivery of growth factors and cells for bone regeneration. Adv. Drug. Deliv. Rev. **64**(12), 1292–1309 (2012)
49. Zeng, F., Lee, H., Allen, C.: Epidermal growth factor-conjugated poly(ethylene glycol)-block-poly(δ-valerolactone) copolymer micelles for targeted delivery of chemotherapeutics. Bioconjugate Chem. **17**(2), 399–409 (2006)
50. Liu, H.W., Chen, C.H., Tsai, C.L., Hsiu, G.H.: Targeted delivery system for juxtacrine signaling growth factor based on rhBMP-2-mediated carrier-protein conjugation. Bone **39**, 825–836 (2006)
51. Izadifar, M., Kelly, M.E., Chen, X.: Regulation of sequential release of growth factors using bilayer polymeric nanoparticles for cardiac tissue engineering. Nanomedicine (Lond) **11**(24), 3237–3259 (2016)
52. Teodori, L., Costa, A., Marzio, R., Perniconi, B., Coletti, D., Adamo, S., Gupta, B., Tarnok, A.: Native extracellular matrix: a new scaffolding platform for repair of damaged muscle. Front Physiol. **5**, 218 (2014)
53. Koh, J., Prabhakar, V., Niklason, L.E.: Decellularized native and engineered arterial scaffolds for transplantation. Cell Transplant. **12**(6), 659–666 (2003)
54. Fishman, J.M., Lowdell, M.W., Urbani, L., Ansari, T., Burns, A.J., Turmaine, M., North, J., Sibbons, P., Seifalian, A.M., Wood, K.J., Birchall, M.A., De Copp, P.: Immunomodulatory effect of a decellularized skeletal muscle scaffold in a discordant xenotransplantation model. Proc. Natl. Acad. Sci. U S A **110**(35), 14360–14365 (2013)
55. Chakraborty, J., Roy, S., Ghos, S.: Regulation of decellularized matrix mediated immune response. Biomater Sci. **8**, 1194–1215 (2020)
56. Sangkert, S., Kamonmattayakul, S., Chai, W.L., Meesane, J.: Modified porous scaffolds of silk fibroin with mimicked microenvironment based on decellularized pulp/fibronectin for designed performance biomaterials in maxillofacial bone defect. Biomed. Mater. Res. Part A **105A**, 1624–1636 (2017)
57. Sangkert, S., Meesane, J., Kamonmattayakul, S., Chai, W.L.: Modified silk fibroin scaffolds with collagen/decellularized pulp for bone tissue engineering in cleft palate: morphological structures and biofunctionalities. Mat. Sci. Eng. C-Mater. **58**, 1138–1149 (2016)
58. Sangkert, S., Kamonmattayakul, S., Chai, W.L., Meesane, J.: A biofunctional-modified silk fibroin scaffold with mimic reconstructed extracellular matrix of decellularized pulp/collagen/fibronectin for bone tissue engineering in alveolar bone resorption. Mater. Lett. **166**, 30–34 (2016)
59. Rajan, N., Habermehl, J., Coté, M.F., Doillon, C.J., Mantovani, D.: Preparation of ready-to-use, storable and reconstituted type I collagen from rat tail tendon for tissue engineering applications. Nat. Protoc. **1**, 2753–2758 (2006)
60. Lu, H., Hoshiba, T., Kawazoe, N., Koda, I., Song, M., Chen, G.: Cultured cell-derived extracellular matrix scaffolds for tissue engineering. Biomaterials **32**, 9658–9666 (2011)
61. Asghari, F., Samiei, M., Adibkia, K., Akbarzadeh, A., Davara, S.: Biodegradable and biocompatible polymers for tissue engineering application: a review. Artif. Cell Nanomed. **45**(2), 185–192 (2017)
62. Zhu, J., Marchant, R.E.: Design properties of hydrogel tissue-engineering scaffolds. Expert Rev. Med. Dev. **8**(5), 607–626 (2011)

# Chapter 3
# Mimicking in Tissue Engineering

Mimicking is an attractive method that has been applied in many areas, for instance, engineering science [1, 2], pharmaceutical science [3], and biomedical science [4]. Mimicking is an approach that has been often been used in innovative designs and for the fabrication of biological systems. For instance, the structure and function of a cocoon provided ideas to create some general items. Generally, mimicking is classified into two main parts: (1) mimicking of structure and (2) function.

Biological systems have been used to create products, based on structural mimicking; an example of this is the sophisticated structure of nature which was used as the model in the design of the building [5, 6]. Another model from the geometric structure of animal, particularly their aerodynamics, has been used to create the outside structure of automobiles [7].

An example of functional mimicking is the repelling of water on the surface of a lotus leaf that was used as a model to design a water-proofing surface [8]. Furthermore, the functionality of a lotus leaf was the inspiration to fabricate some super hydrophobic polymers as coating materials [9]. Super hydrophobic polymers can function in the same fashion as the water proofing of a lotus leaf. Furthermore, the seed-bearing burr that sticks to fabric was the functional model for the invention of VELCRO® [2].

## 3.1 Mimicking in Cell Manipulation

To mimic the structure as well as functionality of cells, the structure of cells has been constructed and then grafted to scaffolds. For instance, the phospholipids on the surface of cells were extracted and grafted onto the polymers [10, 11]. This process was used to mimic the surface function of cells. The grafted polymers were then implanted in the tissue defect. With the mimicked functionality of the cell surface,


J. Meesane, *Mimicked Tissue Engineering Scaffolds for Maxillofacial and Articular Cartilage Surgery*, Engineering Materials,
https://doi.org/10.1007/978-981-19-7830-2_3

the grafted polymer demonstrated suitable biocompatibility [12]. The grafting of phospholipids of cells onto polymers is shown in Fig. 3.1.

Besides the phospholipids on the surface of cells, some structures of the cell surface, which have biofunctionality to enhance tissue regeneration, were grafted onto polymers [13]. In this case, polyethylene glycol (PEG) is a popular polymer used as the base substrate, so as to mimic the biofunctionality of these structures [14]. The process in which to mimic the biofunctionality of those structures, based on

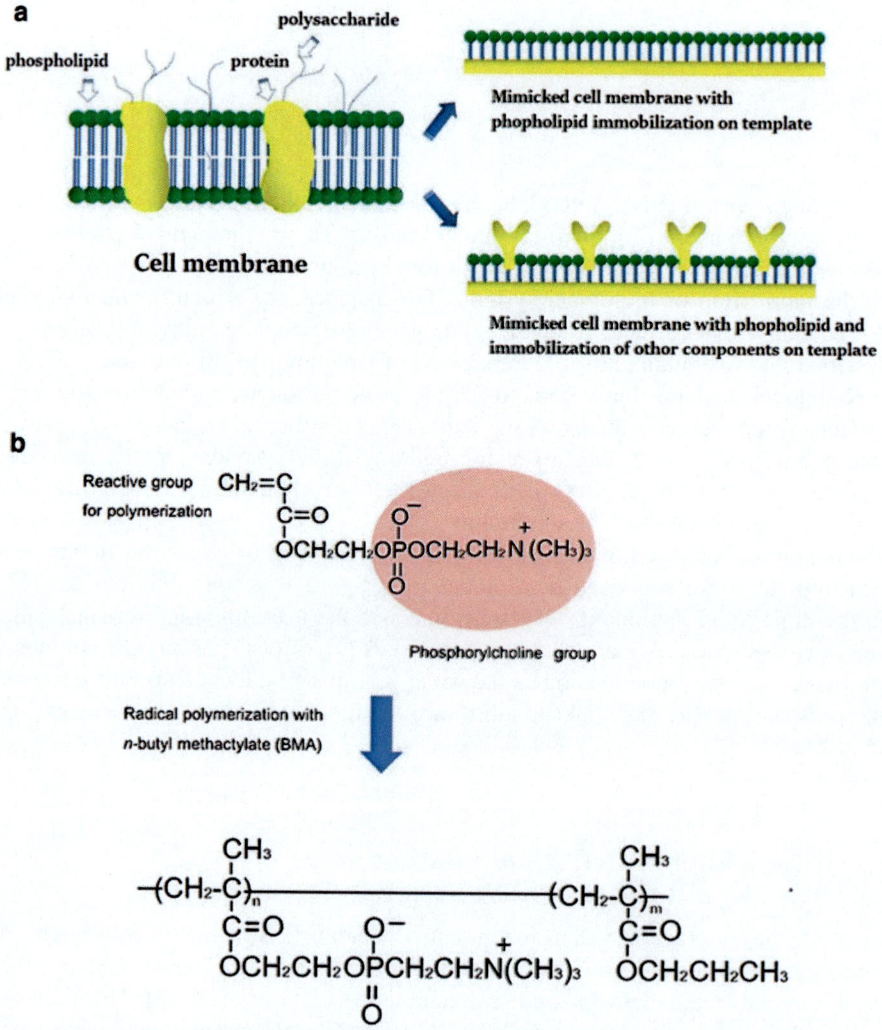


Fig. 3.1 **a** Grafting process of a phospholipid into a polymer and **b** mimicked cell membrane with grafting into polymers

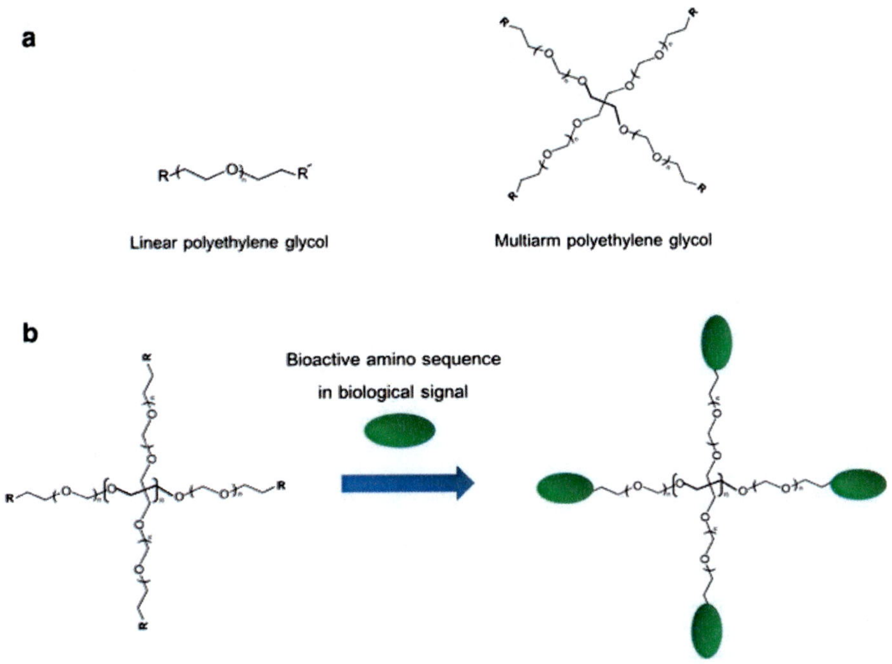


**Fig. 3.2** **a** polyethylene glycol and **b** bioactive amino sequence conjugated with multi-arm polyethylene glycol

PEG, is illustrated in Fig. 3.2. Some research has demonstrated that grafted polymers possess a biological performance to induce tissue regeneration.

## 3.2 Mimicking in the Design of Growth Factors

In the mimicking for the design of growth factors, some research has focused on constructing amino acid sequences in growth factors, for instance, RGD and YIGSR [15, 16]. Amino acid sequences are triggers to enhance tissue regeneration [17]. Principally, the triggers recognize sites for cell adhesion, which in turn leads to the induction of proliferation, migration, and regulation [17]. Adhesion, proliferation, and migration are important factors to promote tissue regeneration [17, 18]. Generally, the amino acid sequences are grafted with polymers that are used via scaffolds and chemical bonding [19]. In order to graft the amino acid sequences in the molecules of polymers, some space molecules have two or more chemical reactive groups [20]. The "di-chemical" reactive groups can react with polymer scaffolds and the amino sequence of growth factors. (Fig. 3.3) [21, 22].

In addition to considering the amino acid sequence of growth factors, some researches were reported on whole molecules of growth factors that were combined

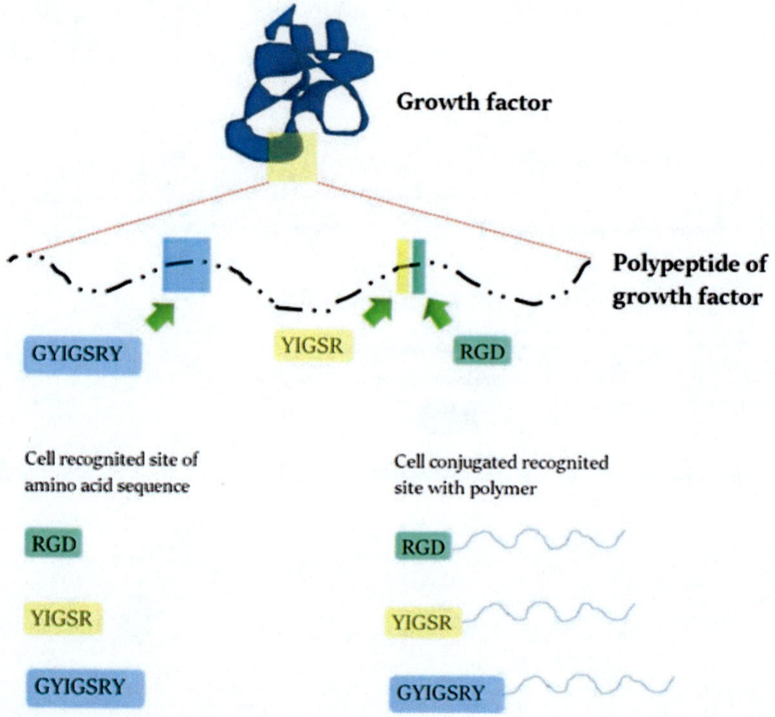


**Fig. 3.3** Conjugation of amino acid sequence in molecules of polymers

with polymers [23, 24]. However, when grafting the growth factors or biomolecules onto polymers, the chemical reactive groups in the growth factors reacted via the spacer molecules (Fig. 3.4) [25]. Normally, the chemical bonding between growth factors and polymers depends on the biological function of the growth factors. It has been reported that chemical bonding could enhance cell behaviors leading to promotion of tissue formation [26]. On the other hand in some cases, chemical bonding would reduce cell behaviors [27]. The differences in cell behavior on the grafted polymer came from the mobility of different growth factor molecules on the polymer molecules [27]. Growth factors had bioactive functionality at the high mobility [27–29]; however, in some cases the growth factors had bioactive functionality at low mobility [27–29].

One simple and popular method to mimic growth factor functionality is the blending of polymers [30]; whereas, growth factors are mixed with polymer solution before fabrication [31]. For some cases, growth factors are encapsulated or form complex structure by amphiphilic polymers [32–34]. This is created for sustainable releasing of growth factors. Amphiphilic polymers demonstrate a physicochemical performance that has both hydrophobic and hydrophilic characteristics. Amphiphilic molecules can form encapsulation with growth factors in spherical structures (Fig. 3.5). The spherical structure has a core–shell organization, with the

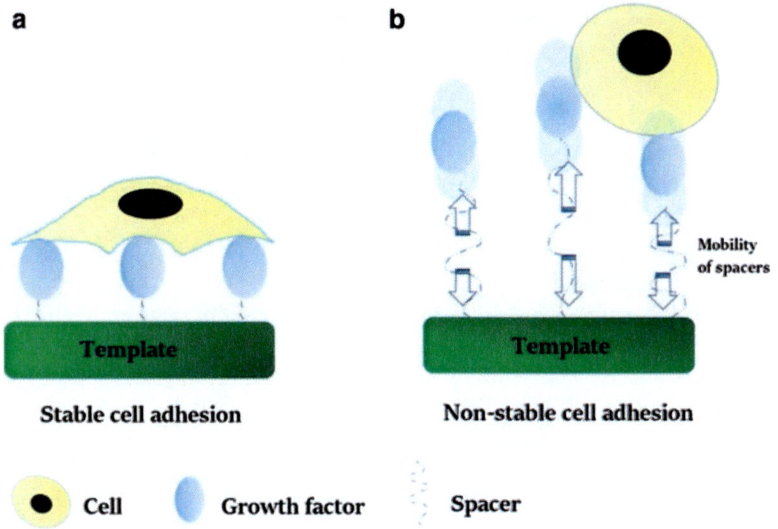


**Fig. 3.4** Behavior of cell adhesion on growth factors grafted templates: **a** stable cell adhesion and **b** non-stable cell adhesion

growth factors in the core. The core often self-organizes by interaction between the growth factor and hydrophobic region of the amphiphilic molecules. In this case, the growth factors must have a hydrophobic region on their molecules that can enhance the interaction. The shell is the hydrophilic region of an amphiphilic molecule that is exposed to the water solution during the mixing.

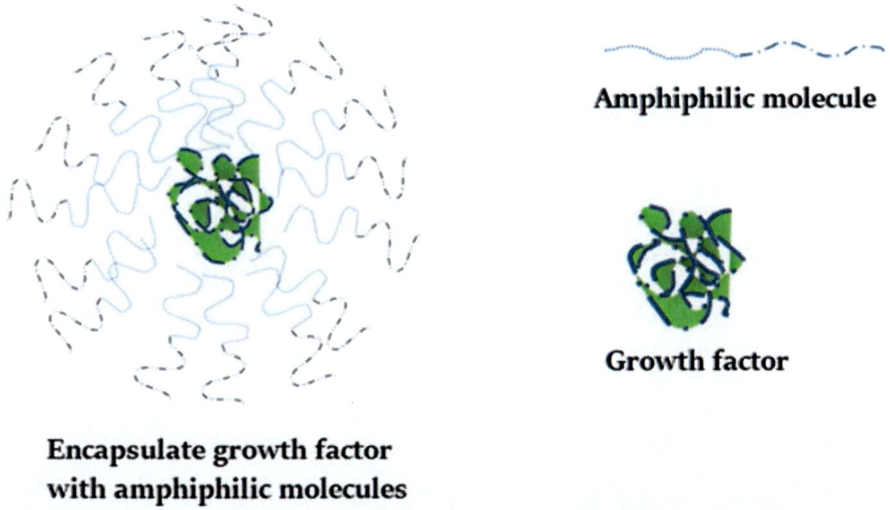


**Fig. 3.5** Formation of encapsulated growth factors by amphiphilic molecules

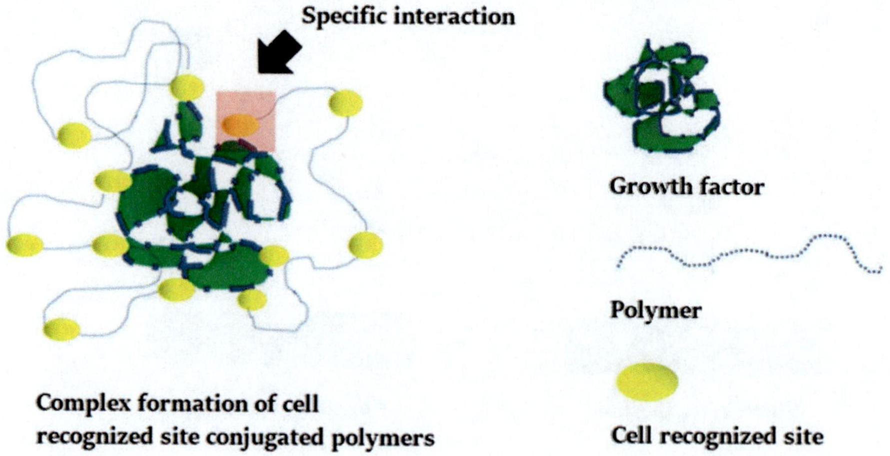


**Fig. 3.6** Self-organization of growth factors with synthesized polymers, via a specific interaction

The formation of a control release system for growth factors is often constructed via physical interactions of hydrophobic and hydrophilic regions, with polymers that have non-specific interactions. Interestingly, in some cases, growth factors are formed via a specific interaction [35]. Specific interactions are generated by polymers that can recognize and attach to the growth factors, upon which the growth factors and polymers then self-organize into a complicated structure (Fig. 3.6).

Based on the principle of self-organization, mimicked growth factor release was constructed to induce tissue regeneration (Fig. 3.7). This is used to create the sequence of growth factor releasing [36, 37]. Therefore, understanding the profile of growth factor release is important in the design to mimic the release of growth factors.

## 3.3 Mimicking in the Design of Extracellular Matrix

Mimicking has often been used in the engineering of scaffolds in three areas of interest: cells, growth factors, and the extracellular matrix or the design of scaffolds. Generally, the aims of mimicking for the engineering of tissue scaffolds are to create a biomaterial that has both the structure and functionality of natural ECM. Therefore, understanding natural ECM is important in the development of suitable scaffolds in tissue engineering. ECM has a complicated network structure of proteins and polysaccharides (Fig. 3.8). In natural ECMs, there are three main components: structural components, connective components, and texture components, and each of these components exhibits different structures and functionalities.

ECM mimicking, in the natural formation of components in tissue, uses some molecules for the construction of a natural structure. The function of structural

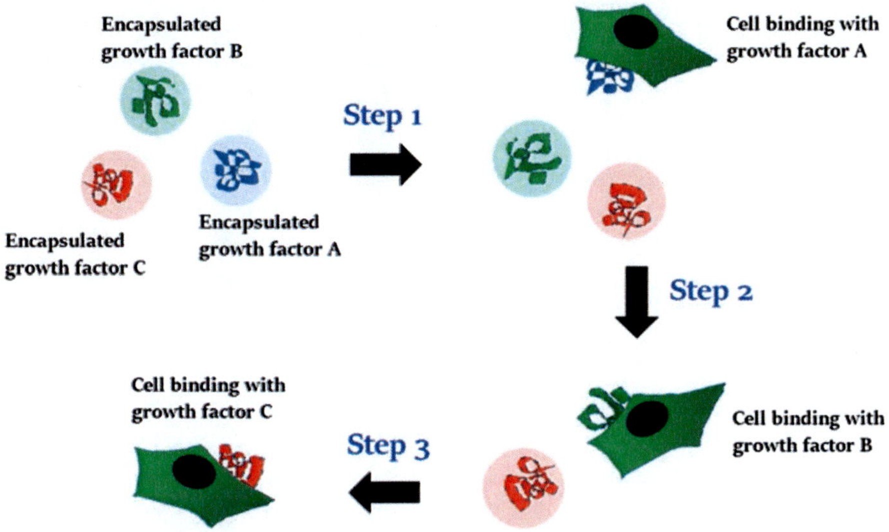


**Fig. 3.7** Mimicking of growth factor release in new tissue formations

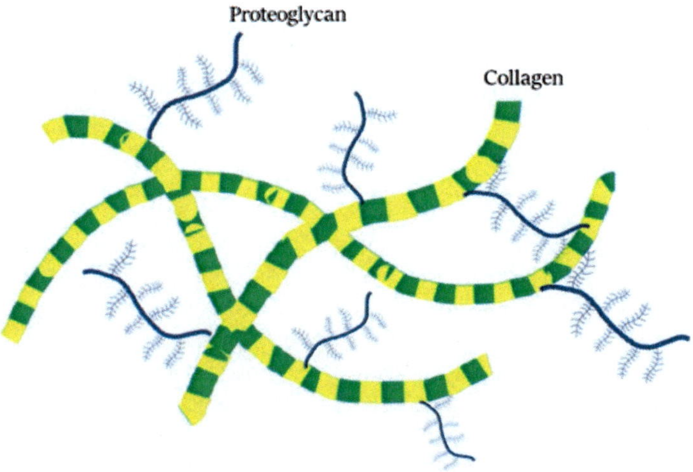


**Fig. 3.8** Structural formation of the extracellular matrix

components is to reinforce the tissue. Dependent on these components, the structures can self-organize into many forms of a fibril structure, which include bundles, dense networks, or loose networks [38, 39]. The organization into these various forms depends on the function of each tissue. Mimicking the forms of the ECM is constructed by some molecules that can self-organize into the natural structure [40].

Self-organization of both natural and synthetic molecules is often used to create an artificial ECM. An artificial ECM has a network fibril structure that comes

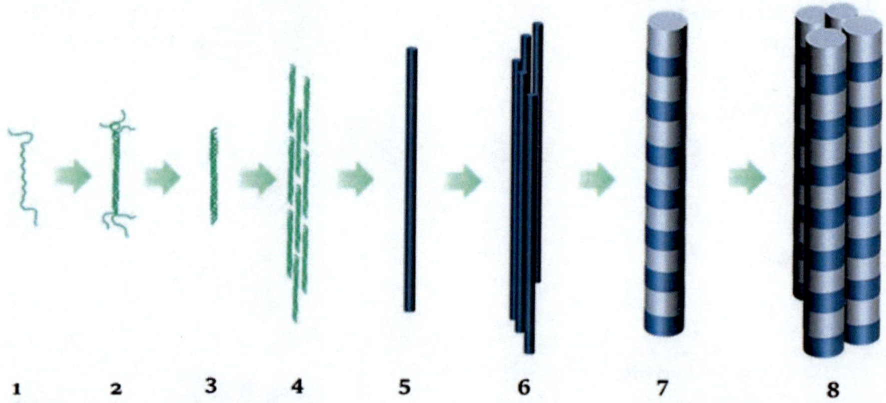


**Fig. 3.9** Assembly of collagen fibers (1) polypeptide of collagen, (2) procollagen or triple helix of polypeptide, (3) tropocollagen or collagen molecule, (4) assembled collagen molecules, (5) collagen fibril, (6) assembled collagen fibrils, (7) collagen fiber, and (8) assembled collagen fibers

from mimicking collagen molecule and its self-assembly [41–43]. The assembly of network fibrils, in natural tissue, is constructed by collagen molecules that are in the nanoscale size range. Collagen molecules are secreted from cells into the extracellular matrix; then these molecules start to form seeding clusters. These clusters then grow into microfibrils that act as nucleators of fibers, which in turn then grow into fibers (Fig. 3.9) [44, 45].

The mimicking of collagen fibrils is prepared by collagen molecules dissolved in a mildly acidic solution. This mildly acidic solution exhibits a low deconstruction of collagen molecules [46]. For this, acetic acid is often used as the acidic solution, which also exhibits the salting out phenomenon [47]. The salting out phenomenon generates ions in the solution. Those ions can pull out water molecules that are caged around the collagen molecules [47]. Then, the collagen molecules can organize themselves into a regular structure, via molecular interaction [46, 47].

There are many ions that show the strength of salting out according to the Hofmeiter series [47] (Fig. 3.10). Howbeit, when the collagen molecules are dissolved in an acidic solution that has the salting in phenomenon, the water molecules are maintained around the collagen molecules. The salting in phenomenon causes rapid denaturation of the separated collagen molecules by the water molecules. Therefore, mimicking collagen into a fibril structure requires a mildly acidic solution that has the salting out phenomenon.

After the collagen molecules are dissolved in an acidic solution, they are neutralized by a buffer solution [48, 49]. In this neutralized solution, collagen molecules can assemble into a fibril structure. Collagen molecules are sensitive to both pH and temperature [49]; thus, the pH should not be much lower or much higher than 7; otherwise, the collagen molecules will be disturbed causing the assembly into the fibril structure to be retarded. Assembly in either higher or lower pH ranges always results in irregular fibril structures [49]. Collagen molecules can assemble into a

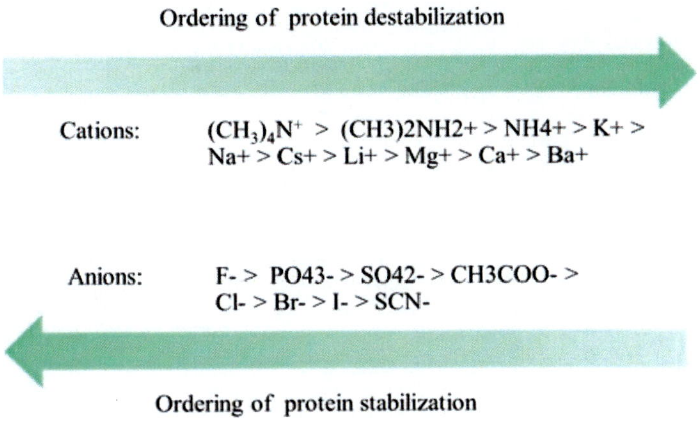


**Fig. 3.10** Hofmeiter series of salting out and salting in

regular structure at a temperature around 37 °C [49]. At temperatures lower than 37 °C, assembly of the collagen molecules is retarded and fibril formation is slow. At temperatures higher than 37 °C, assembly of collagen is disturbed, and in some cases, the collagen molecules are denatured. Collagen can form fibril structures at a pH of around 7 and a temperature of 37 °C. Besides pH and temperature, concentration also has an effect on collagen assembly [50]. Therefore, the preparation to mimic collagen fibril structures, as natural structures, in the ECM requires good control of these conditions.

ECM mimicking of the natural structure in tissue is often constructed of collagen molecules that have been often combined with other molecules in the ECM, these being either natural or synthetic molecules. Natural and synthetic molecules can function biologically in the same manner as components in natural ECM. For instance, collagen molecules can combine with hyaluronic acid, which has the important role of inducing cell migration [51]. Collagens have some domains in the molecules that can form a collagen/hyaluronic acid interaction, leading to a sophisticated structural organization [51]. Furthermore, polysaccharides, such as alginate, chitosan, cellulose derivatives, and glycosaminoglycans, can be added to the collagen molecules [52–54]. The combination of polysaccharides with collagen will result in a self-organized network structure of collagen fibers embedded in a hydrogel matrix (Fig. 3.11). These combined molecules can then be used to design a mimicked ECM similar to natural tissue.

The formation of combined collagen molecules, with these biological molecules, begins with molecular interaction with the collagen molecules. The molecular interaction of each biological molecule demonstrates different characteristics that disturb fibril formation [55]. Some molecules reduce the size of the collagen fibrils [55], whereas some molecules induce larger collagen fibrils [55]. Although biological molecules can enhance the fibril packing into a dense network structure, in some

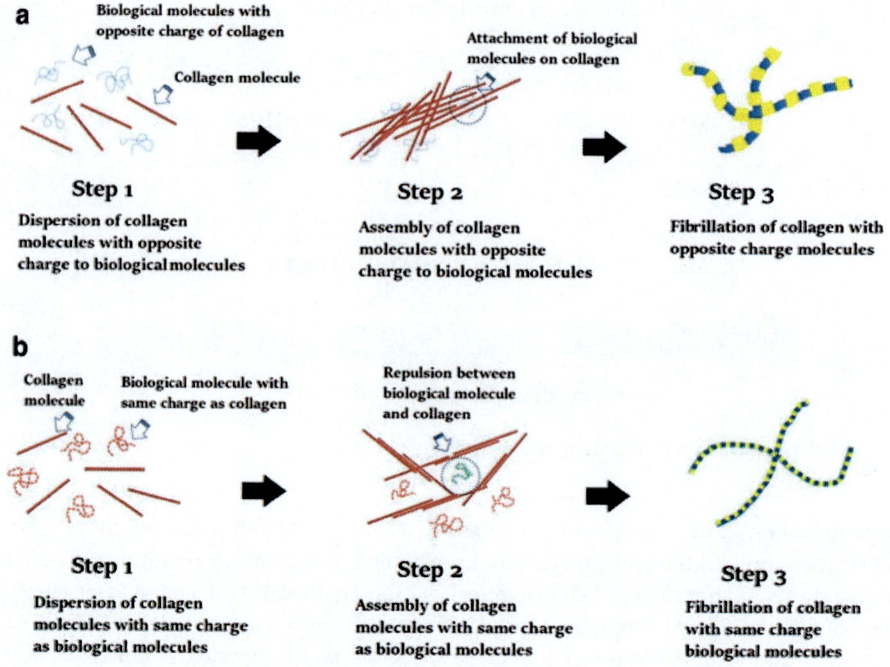


**Fig. 3.11** Assembly of collagen molecules with biological molecules

cases the biological molecules can reduce fibril packing, leading to the formation of a loose network structure [52, 55].

Modification of collagen molecules into various structures is often used to create various structures of collagen assembly [56]. These various structures are constructed by connecting to the recognized domain or molecules that have organized themselves into sophisticated structures [57]. Sophisticated structures such as these exhibit various biological functions [57]. Elastin is another material that has been used for assembly of fibril networks, in kin to natural structures [58].

In some research, mimicked structural ECMs were constructed by oligo-peptides that were prepared by synthesis into hydrophobic and hydrophilic regions [59, 60]. Both of the hydrophobic and hydrophilic regions can organize into fibril structures similar to collagen type I [60]. When the structure of oligo-peptides was modified, the assembly of molecules changed. Those modified oligo-peptide molecules self-organized into a network structure similar to some loose tissues [61, 62]. Oligo-peptide molecules are often grafted with biofunctional molecules to induce cell adhesion, proliferation, and migration. The grafted oligo-peptide molecules show mimicked functionality, similar to a natural extracellular matrix that can enhance tissue regeneration [63, 64]. Normally, whole biofunctional molecules of proteins are grafted onto oligo-peptide molecules. In some cases, the sequences of amino acids of proteins in the extracellular matrix were selected and then grafted onto the oligo-peptides [65]. These sequences of amino acids can be recognized for cell adhesion, proliferation, and regulation [66]. Therefore, the sequences of amino acids are important in biological performance, so as to induce cell adhesion, proliferation, and regulation, which leads to the enhancement of tissue regeneration.

In addition to oligo-peptides and their derivatives, some polymers were modified by combination with biofunctional molecules, and this was shown to trigger the induction of tissue regeneration. For this, a chemical reaction was used to combine said molecules. Biofunctional molecules that often combine with polymers are cell receptors, enzyme molecules as well as some sequenced amino acids of the extracellular matrix. Generally, multi-arms PEG is selected as the main structure for combination with biofunctional molecules [67, 68]. When the modified multi-arms PEGs were implanted into the tissue defect, the combined biofunctional molecules acted as signals to regulate the cells in the proposed tissue. Therefore, combining the trigger molecules is important to mimic the biofunctional performance of scaffolds as natural tissue.

## 3.4 Mimicking in the Design of the Microenvironment

The microenvironment system is a combination of growth factors or soluble signals and the ECM or insoluble signals [69, 70]. Those components self-organize into a sophisticated structure and functionality. One simple method to prepare a microenvironment is the extraction from natural tissue of so-called decellularized tissue. Decellularized tissue has problems with fragments of ECM along with other signals during cell isolation [71]. Therefore, some researchers use certain molecules to combine with those fragments, so as to repair the structure and functionality, similar to natural microenvironments (Fig. 3.12). In this case, fibronectin is able to connect the molecules between the ECM and growth factors [71].

Furthermore, some molecules are used as crosslinked ECM incorporated with other signals [72, 73]. In application, decellularized tissue has been used as scaffolds to enhance the potential of tissue regeneration [74]. Furthermore, in some research, decellularized tissue was dissolved into solution before coating onto scaffolds [70, 71].

In some cases, polymers have been used as the main structure to combine trigger molecules of growth factors and the extracellular matrix. This is the method to create a microenvironment for different tissues. With the use of this method, the microenvironment can be created into programs of cell regulation, by conjugated polymers with the sequence of the trigger molecules during tissue regeneration. This can enhance the potential of tissue regeneration (Fig. 3.13) [70, 75, 76].

Additionally, some trigger molecules, or amino acid sequences that can be degraded by enzymes, are combined with [77, 78]. This can create biodegradable polymers that perform as a microenvironment in natural tissue. Generally, microenvironments are digested by enzymes that can promote degraded spaces. Therefore,

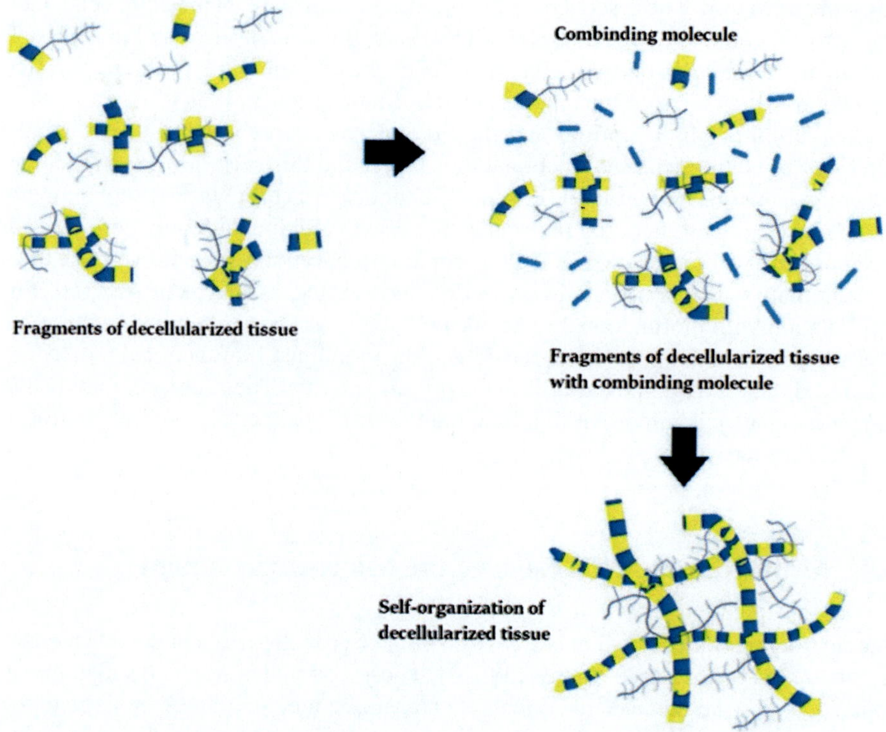


**Fig. 3.12** Self-organization of repaired decellularized tissue with combining molecules

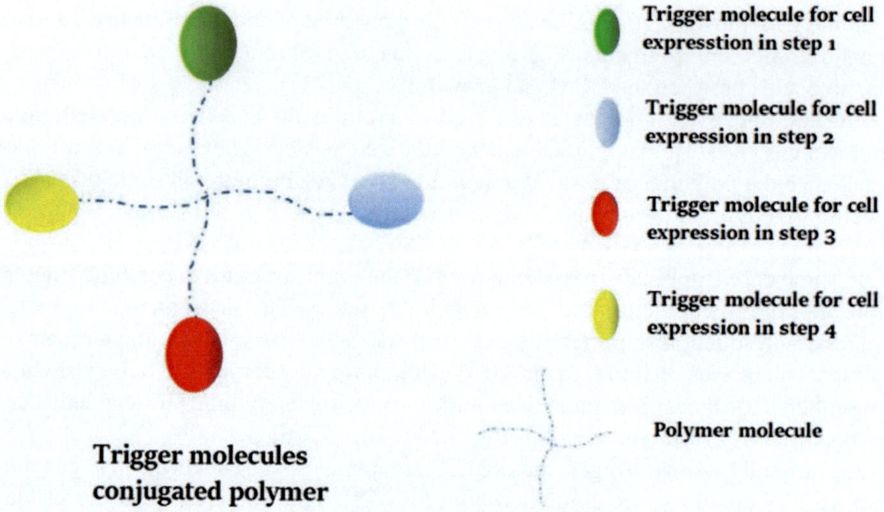


**Fig. 3.13** Conjugated polymers for cell regulation programming

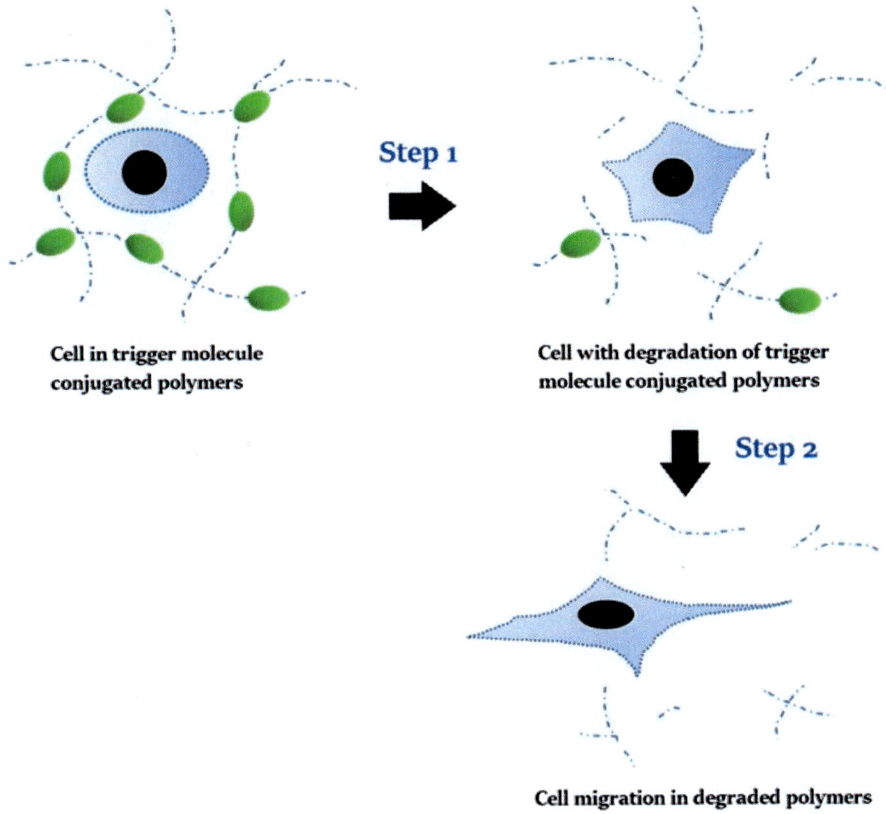


**Fig. 3.14** Cell migration in a degraded space of polymers, conjugated with trigger molecules

a degraded space can promote cell migration, leading to the enhancement of tissue regeneration (Fig. 3.14).

In summary, the mimic approach is classified into four routes: (1) cells, (2) growth factors, (3) the extracellular matrix, and (4) microenvironments. An understanding of the mimic approach is necessary before construction or fabrication of high performance scaffolds for tissue regeneration.

## References


1. Benyus, J.M.: Biomimicry: innovation inspired by nature. Perennial, New York (2002)
2. Hwang, J., Jeong, Y., Park, J.M., Lee, K.H., Hong, J.W., Choi, J.: Biomimetics: forecasting the future of science, engineering, and medicine. Int. J. Nanomedicine **10**, 5701–5713 (2015)
3. Sabu, C., Rejo, C., Kotta, S., Pramod, K.: Bioinspired and biomimetic systems for advanced drug and gene delivery. J. Control Release **287**, 142–155 (2018)
4. Zhang, G.: Biomimicry in biomedical research. Organogenesis **8**(4), 101–102 (2012)

5. Knippers, J., Speck, T.: Design and construction principles in nature and architecture. Bioinspir. Biomim. **7**, 015002 (2012)
6. Aziz, M.S., El Sheriff, A.Y.: Biomimicry as an approach for bio-inspired structure with the aid of computation. Alex Eng. J. **55**(1), 707–714 (2016)
7. Wijegunawardana, I.D., de Mel, W.R.: Biomimetic designs for automobile engineering: a review. Int. J. Auto Mech. Eng. **18**(3), 9029–9041 (2021)
8. Collins, C.M., Safiuddin, M.D.: Lotus-leaf-inspired biomimetic coatings: different types, key properties, and applications in infrastructures. Infrastructure **7**(4), 46 (2022)
9. Latthe, S.S., Terashima, C., Nakata, K., Fujishima, A.: Superhydrophobic surfaces developed by mimicking hierarchical surface morphology of lotus leaf. Molecules **19**(4), 4256–4283 (2014)
10. Cashion, M.P., Long, T.E.: Biomimetic design and performance of polymerizable lipids. Acc. Chem. Res. **42**(8), 1016–1025 (2009)
11. Ishihara, K., Goto, Y., Takai, M., Matsuno, R., Inoue, Y., Konno, T.: Novel polymer biomaterials and interfaces inspired from cell membrane functions. Biochim. Biophys. Acta. **1810**(3), 268–275 (2011)
12. Ishihara, K.: Bioinspired phospholipid polymer biomaterials for making high performance artificial organs. Sci. Tech. Adv. Mat. **1**(3), 131–138
13. Zhao, J., Santino, F., Giacomini, D., Gentilucci, L.: Integrin-targeting peptides for the design of functional cell-responsive biomaterials. Biomedicines **8**(9), 307 (2020)
14. Jabbari, E.: Bioconjugation of hydrogels for tissue engineering. Curr. Opin. Biotechnol. **22**(5), 655–660 (2011)
15. Hersel, U., Dahmen, C., Kessler, H.: RGD modified polymers: biomaterials for stimulated cell adhesion and beyond. Biomaterials **24**, 4385–4415 (2003)
16. Smith, L.A., Sibai, C., Barker, X.A., Zheng, J., Reneker, D.H., Dove, A.P., Becker, M.L.: Directed differentiation and neurite extension of mouse embryonic stem cell on aligned poly(lactide) nanofibers functionalized with YIGSR peptide. Biomaterials **34**, 9089–9095 (2013)
17. Hosoyama, K., Lazurko, C., Muñoz, M., McTiernan, C.D., Alarcon, E.I.: Peptide-based functional biomaterials for soft-tissue repair. Front Bioeng. Biotechnol. **7**, 205 (2019)
18. Orcid, Y.L., Xiao, Y., Liu, C.: The horizon of materiobiology: a perspective on material-guided cell behaviors and tissue engineering. Chem. Rev. **117**(5), 4376–4421 (2017)
19. Medeiros BorsagliI, F.G.L., Carvalho, I.C., Mansur, H.S.: Amino acid-grafted and N-acylated chitosan thiomers: Construction of 3D bio-scaffolds for potential cartilage repair applications. Int. J. Biol. Macromol. **114**, 270–282 (2018)
20. Hamley, I.W.: PEG–Peptide Conjugates. Biomacromol. **15**(5), 1543–1559 (2014)
21. Sharon, J.L., Puleo, D.A.: Immobilization of glycoproteins, such as VEGF, on biodegradable substrates. Acta Biomater. **4**, 1016–1023 (2008)
22. Masters, K.S.: Covalent growth factor immobilization strategies for tissue repair and regeneration. Macromol. Biosci. **11**, 1149–1163 (2011)
23. Janicki, P., Schmidmaier, G.: What should be the characteristics of the ideal bone graft substitute? Combining scaffolds with growth factors and/or stem cells. Injury **42**, S77–S81 (2011)
24. Hajimiri, M., Shahverdi, S., Kamalinia, G., Dinarvand, R.: Growth factor conjugation: strategies and applications. J. Biomed. Mater. Res. A **103**, 819–838 (2015)
25. Smith, S., Goodge, K., Delaney, M., Struzyk, A., Tansey, N., Frey, M.A.: Comprehensive review of the covalent immobilization of biomolecules onto electrospun nanofibers. Nanomaterials-Basel **10**, 2142 (2020)
26. Gomeza, N., Lu, Y., Chen, S., Schmidt, C.E.: Immobilized nerve growth factor and microtopography have distinct effects on polarization versus axon elongation in hippocampal cells in culture. Biomaterials **28**, 271–284 (2007)
27. Psarra, E., Foster, E., König, U., You, J., Ueda, Y., Eichhorn, K.J., Müller, M., Stamm, M., Revzin, A., Uhlmann, P.: Growth factor-bearing polymer brushes—Versatile bioactive substrates influencing cell response. Biomacromol. **16**(11), 3530–3542 (2015)

28. Pompe, T., Salchert, K., Alberti, K., Zandstra, P., Werner, C.: Immobilization of growth factors on solid supports for the modulation of stem cell fate. Nat. Protoc. **5**, 1042–1050 (2010)
29. Ham, T.R., Farrag, M., Leipzig, N.D.: Covalent growth factor tethering to direct neural stem cell differentiation and self-organization. Acta Biomater. **53**, 140–151 (2017)
30. Shen, H., Hu, X.: Growth factor loading on aliphatic polyester scaffolds. RSC Adv. **11**, 6735–6747 (2021)
31. Ji, W., Sun, Y., Yang, F., van den Beucken, J.J.J.P., Fan, M., Chen, Z., Jansen, J.A.: Bioactive electrospun scaffolds delivering growth factors and genes for tissue engineering applications. Pharm. Res-Dordr **28**, 1259–1272 (2011)
32. Wen, Y., Li, F., Li, C., Yin, Y., Li, J.: High mechanical strength chitosan-based hydrogels cross-linked with poly(ethylene glycol)/polycaprolactone micelles for the controlled release of drugs/growth factors. J. Mater. Chem. B **5**, 961–971 (2017)
33. Sukarto, B., Amsden, B.G.: Low melting point amphiphilic microspheres for delivery of bone morphogenetic protein-6 and transforming growth factor-β3 in a hydrogel matrix. J. Control. Release. **158**, 53–62 (2012)
34. Hosseinkhani, H., Hosseinkhani, M., Khademhosseini, A., Kobayashi, H., Tabata, Y.: Enhanced angiogenesis through controlled release of basic fibroblast growth factor from peptide amphiphile for tissue regeneration. Biomaterials **27**, 5836–5844 (2006)
35. Joung, Y.K., Bae, J.W., Park, K.D.: Controlled release of heparin-binding growth factors using heparin-containing particulate systems for tissue regeneration. Expert Opin. Drug Del. **5**, 1173–1184 (2008)
36. Min, Q., Liu, J., Yu, X., Zhang, Y., Wu, J., Wan, Y.: Sequential delivery of dual growth factors from injectable chitosan-based composite hydrogels. Mar. Drugs **17**(6), 365 (2019)
37. Lee, K., Silva, E.A., Mooney, D.J.: Growth factor delivery-based tissue engineering: general approaches and a review of recent developments. J. R. Soc. Interface **8**(55), 153–170 (2011)
38. Theocharis, A.D., Skandalis, S.S., Gialelia, C., Karamanos, N.K.: Extracellular matrix structure. Adv. Drug Deliver Rev. **97**, 4–27 (2016)
39. Mouw, J.K., Ou, G., Weaver, V.M.: Extracellular matrix assembly: a multiscale deconstruction. Nat. Rev. Mol. Cell Bio. **15**, 771–785 (2014)
40. Sevilla, C.A., Dalecki, D., Hocking, D.C.: Extracellular matrix fibronectin stimulates the self-assembly of microtissues on native collagen gels. Tissue Eng. Pt A **16**, 3805–3819 (2010)
41. Luo, J., Tong, Y.W.: Self-assembly of collagen-mimetic peptide amphiphiles into biofunctional nanofiber. ACS Nano. **5**(10), 7739–7747 (2011)
42. Fallas, J.A., O'Learya, L.E.R., Hartgerink, J.D.: Synthetic collagen mimics: self-assembly of homotrimers, heterotrimers and higher order structures. Chem. Soc. Rev. **39**, 3510–3527 (2010)
43. Kotch, F.W., Raines, R.T.: Self-assembly of synthetic collagen triple helices. PNAS **103**(9), 3028–3033 (2006)
44. Zhu, S., Yuan, Q., Yin, T., You, J., Gu, Z., Xiong, S., Hu, Y.: Self-assembly of collagen-based biomaterials: preparation, characterizations and biomedical applications. J. Mater. Chem. B **6**, 2650–2676 (2018)
45. Cisneros, D.A., Hung, C., Franz, C.M., Muller, D.J.: Observing growth steps of collagen self-assembly by time-lapse high-resolution atomic force microscopy. J. Struct. Biol. **154**, 232–245 (2006)
46. Caputo, I., Lepretti, M., Scarabino, C., Esposito, C., Proto, A.: An acetic acid-based extraction method to obtain high quality collagen from archeological bone remains. Anal. Biochem. **421**, 92–96 (2012)
47. Hyde, A.M., Zultanski, S.L., Waldman, J.H., Zhong, Y.L., Shevlin, M., Peng, F.: General principles and strategies for salting-out informed by the hofmeister series. Org. Process Res. Dev. **21**(9), 1355–1370 (2017)
48. Puttawibul, P., Benjakul, S., Meesane, J.: An in situ hydrogel of mimicked self-assembly type I collagen from shark skin (brownbanded bamboo shark, Chiloscyllium punctatum)/methylcellulose for central nerve system regeneration: preparation and characterization. J. Biomim. Biomater. Biomed. Eng. **14**, 14–29 (2015)

49. González-Masís, J., Cubero-Sesin, J.M., Guerrero, S., González-Camacho, S., Corrales-Ureña, Y.R., Redondo-Gómez, C., Vega-Baudrit, J.R., Gonzalez-Paz, R.J.: Self-assembly study of type I collagen extracted from male Wistar Hannover rat tail tendons. Biomater. Res. **24**(19) (2020)
50. Yana, M., Li, B., Zhao, X., Qin, S.: Effect of concentration, pH and ionic strength on the kinetic self-assembly of acid-soluble collagen from walleye pollock (Theragra chalcogramma) skin. Food Hydrocolloid **29**, 199–204 (2012)
51. Fujie, T., Furutate, S., Niwa, D., Takeok, S.: A nano-fibrous assembly of collagen–hyaluronic acid for controlling cell-adhesive properties. Soft Matt. **6**, 4672–4676 (2010)
52. Tsai, S.W., Liu, R.L., Hsu, F.Y., Chen, C.C.: A study of the influence of polysaccharides on collagen self-assembly: Nanostructure and kinetics. Biopolymer **83**, 381–388 (2006)
53. Salchert, K., Oswald, J., Streller, U., Grimmer, M., Herold, N., Werner, C.: Fibrillar collagen assembled in the presence of glycosaminoglycans to constitute bioartificial stem cell niches in vitro. J. Mater. Sci. Mater. Med. **16**(6), 581–585 (2005)
54. Rühland, C., Schönherr, E., Robenek, H., Hansen, U., Iozzo, R.V., Bruckner, P., Seidler, D.G.: The glycosaminoglycan chain of decorin plays an important role in collagen fibril formation at the early stages of fibrillogenesis. FEBS J. **274**, 4246–4255 (2007)
55. Sant, S., Coutinho, D.F., Gaharwar, A.K., Neves, N.M., Reis, R.L., Gomes, M.E., Khademhosseini, A.: Self-assembled hydrogel fiber bundles from oppositely charged polyelectrolytes mimic micro-/nanoscale hierarchy of collagen. Adv. Funct. Mater. **27**, 1606273 (2017)
56. Zhang, J., Tu, X., Wang, W., Nan, J., Wei, B., Xu, C., He, L., Xu, Y., Wang, H.: Insight into the role of grafting density in the self-assembly of acrylic acid-grafted-collagen. Int. J. Biol. Macromol. **128**, 885–892 (2019)
57. Birk, D.E.: Type V collagen: heterotypic type I/V collagen interactions in the regulation of fibril assembly. Micron **32**, 223–237 (2001)
58. Wang, R., Ozsvar, J., Yeo, G.C., Weiss, A.S.: Hierarchical assembly of elastin materials. Curr. Opin. Chem. Eng. **24**, 54–60 (2019)
59. Fung, S.Y., Keyes, C., Duhamel, J., Chen, P.: Concentration effect on the aggregation of a self-assembling oligopeptide. Biophys. J. **85**, 537–548 (2003)
60. Hamley, I.W., Dehsorkhi, A., Castelletto, V., Seitsonen, J., Ruokolainen, J., Iatrou, H.: Self-assembly of a model amphiphilic oligopeptide incorporating an arginine headgroup. Soft Matt. **9**, 4794–4801 (2013)
61. Jesus, C.N., Evans, R., Forth, J., Estarellas, C., Gervasio, F.L., Giuseppe, B.G.: Amphiphilic histidine-based oligopeptides exhibit pH-reversible fibril formation. ACS Macro. Lett. **10**(8), 984–989 (2021)
62. Koga, T., Higuchi, M., Kinoshita, T., Nobuyuki, H.N.: Controlled self-assembly of amphiphilic oligopeptides into shape-specific nanoarchitectures. Chem. Eur. J. **12**, 1360–1367 (2006)
63. Levin, A., Hakala, T.A., Schnaider, L., Bernardes, G.J.L., Gazit, E., Knowles, T.P.J.: Biomimetic peptide self-assembly for functional materials. Nat. Rev. Chem. **4**, 615–634 (2020)
64. Fichman, G., Gazit, E.: Self-assembly of short peptides to form hydrogels: Design of building blocks, physical properties and technological applications. Acta Biomater. **10**, 1671–1682 (2014)
65. Sun, Y., Li, W., Wu, X., Zhang, N., Zhang, Y., Ouyang, S., Song, X., Fang, X., Seeram, R., Xue, W., He, L., Wutian, Wu.W.: Functional self-assembling peptide nanofiber hydrogels designed for nerve degeneration. ACS Appl. Mater. Interfaces **8**(3), 2348–2359 (2016)
66. Barker, T.H.: The role of ECM proteins and protein fragments in guiding cell behavior in regenerative medicine. Biomaterial **32**, 4211–4214 (2011)
67. Pasut, G., Veronese, F.M.: PEG conjugates in clinical development or use as anticancer agents: an overview. Adv. Drug. Deliver Rev. **61**, 1177–1188 (2009)
68. Qi, Y., Chilkoti, A.: Protein–polymer conjugation—moving beyond PEGylation. Curr. Opin. Chem. Biol. **28**, 181–193 (2015)
69. Rijal, G., Li, W.: Native-mimicking in vitro microenvironment: an elusive and seductive future for tumor modeling and tissue engineering. J. Biol. Eng. **12**, 20 (2018)
70. Sangkert, S., Meesane, J., Kamonmattayakul, S., Chai, W.L.: Modified silk fibroin scaffolds with collagen/decellularized pulp for bone tissue engineering. Mat. Sci. Eng. C-Mater. **58**, 1138–1149 (2016)

71. Sangkert, S., Kamonmattayakul, S., Chai, W.L., Meesane, J.: Modified porous scaffolds of silk fibroin with mimicked microenvironment based on decellularized pulp/fibronectin for designed performance biomaterials in maxillofacial bone defect. J. Biomed. Mater. Res. A **105**, 1624–1636 (2017)
72. Liu, Y., Cai, S., Shu, X.Z., Shelby, J., Prestwich, G.D.: Release of basic fibroblast growth factor from a crosslinked glycosaminoglycan hydrogel promotes wound healing. Wound Repair Regen **15**, 245–251 (2007)
73. Elia, R., Fuegy, P.W., VanDelden, A., Firpo, M.A., Prestwich, G.D., Peattie, R.A.: Stimulation of in vivo angiogenesis by in situ crosslinked, dual growth factor-loaded, glycosaminoglycan hydrogels. Biomaterial **31**, 4630–4638 (2010)
74. Cheng, C.W., Solorio, L.D., Alsberg, E.: Decellularized tissue and cell-derived extracellular matrices as scaffolds for orthopaedic tissue engineering. Biotechnol Adv. **32**, 462–484 (2014)
75. Tsurkan, M.V., Chwalek, K., Prokoph, S., Zieris, A., Levental, K.R., Freudenberg, U., Werner, C.: Defined polymer-peptide conjugates to form cell-instructive starPEG–heparin matrices in situ. Adv. Mater. **25**, 2606–2610 (2013)
76. Guoa, C., Kim, H., Ovadia, E.M., Mourafetis, C.M., Yang, M., Chen, W., Kloxin, A.M.: Bio-orthogonal conjugation and enzymatically triggered release of proteins within multi-layered hydrogels. layered hydrogels. Acta Biomaterialia **56**, 80–90 (2017)
77. Fonseca, K.B., Maia, F.R., Cruz, F.A., Andrade, D., Juliano, M.A., Granja, P.L., Cristina, C., Barrias, C.C.: Enzymatic, physicochemical and biological properties of MMP-sensitive alginate hydrogels. Soft Matt. **9**, 3283–3292 (2013)
78. Chau, Y., Tan, F.E., Robert, L.R.: Synthesis and characterization of dextran−peptide−methotrexate conjugates for tumor targeting via mediation by matrix metalloproteinase II and matrix metalloproteinase IX. Bioconjugate Chem. **15**(4), 931–941 (2004)

# Chapter 4
# Mimicked Molecular Structures in Scaffolds

Scaffolds are important in tissue engineering as they maintain the shape of tissue during regeneration. Therefore, structural mimicking of scaffolds, similar to natural tissue, can enhance the performance of tissue formation [1, 2]. The extracellular matrix (ECM), which is the base component in natural tissue, is often used as the model for guidance of mimicked structure in scaffolds. ECM exhibits self-organization into different, sophisticated structures, depending on the types of tissue.

Some tissues need a loose network structure fitting to the physical and mechanical properties of the tissue, for instance skin, adipose, the liver, and cartilage [2–5]. These tissues require a large amount of nutrients for support during formation [1, 6]. Therefore, the design of the scaffold for these tissues needs to be a loose structure, with large pore sizes, for supporting diffusion of nutrients [1, 6].

In contrast to this, some tissues need a dense structure that is able to maintain the shape and support the mechanical function like natural tissue, such as ligaments, meniscus, and bones [7–9]. These scaffolds need a regular structure that has high force resistance [7–9]. In the case of ligaments, ECMs have a regular orientation of collagen fibers. Scaffolds such as these are often constructed into knits or braid that have a highly ordered structure [7, 10]. A highly ordered structure of fibers can resist tensile forces, as in the native biomechanics of ligament tissue.

In the case of meniscus tissue, some scaffolds are designed into a regular orientation of natural or synthetic fibers [11, 12]. This kind of structural orientation can support the biomechanical loads similar to native meniscus tissue [13].

The structure of bone tissue are organized as composite of both collagen and hydroxyapatite. Inside the structure of fibers are nano-plates, and this structure has both high strength and toughness. Scaffolds are often constructed by assembly of collagen fibrils and the biomineralization of calcium phosphate, based on the mimic concept [14].

In this chapter, mimicking in the structure of scaffolds, as natural ECMs, is emphasized in the design of high performance tissue regeneration and contains four main



J. Meesane, *Mimicked Tissue Engineering Scaffolds for Maxillofacial and Articular Cartilage Surgery*, Engineering Materials,
https://doi.org/10.1007/978-981-19-7830-2_4

sections: (1) mimicked molecular structure of ECMs in scaffolds, (2) mimicked self-organization based on synthetic molecules, (3) molecular modification in mimicked molecular structures, and (4) chemical treatment for mimicked molecular structures.

## 4.1 Mimicked Molecular Structure of ECM in Scaffolds

Generally, the molecular structure of the ECM plays an important role, which has an effect on tissue formation. For instance, the molecular structure of some proteins containing amino acid motif of RGD as cell recognized binding sites shows the performance to promote cell behaviors for enhancement of tissue formation. Amino acids in molecule makeup of proteins have influence on the regularity of chain conformations: random coils, alpha-helix, and beta-sheet. These conformations have an effect on cell behavior for tissue formation. In the case of polysaccharide in ECMs, some monosaccharides show the performance as a binding site for their connection to proteins. This leads to the complicated formation in molecular structures which has an influence to promote cell behaviors for enhancement of tissue formation.

### *4.1.1 Mimicked Amino Acid Motifs and the Molecular Chain of Proteins*

Amino motifs and the molecular chain of proteins have important roles, acting as binding sites for cell attachment [15–17]. This leads to promoting other cell behaviors, for instance, proliferation, migration, and differentiation [18]. To mimic amino acid motif and the molecular chain of proteins, they are first synthesized (Fig. 4.1), and then, they are directly fabricated into scaffolds. An example of this is synthesized molecular chains of proteins which are fabricated into hydrogel scaffolds [19]. In the case of amino acid motifs, they need to incorporate with natural or synthetic polymers by either molecular grafting, conjugation, or surface immobilization [20–22]. For instance, RGD is grafted on the side chains of polyethylene glycol before being fabricated into hydrogel scaffolds [23]. Normally, amino acid motifs, RGD and YIGSR, are often used to mimic molecular structures [24, 25]. For mimicking, for some cases, the dual amino acid motifs or molecular chains are selected to create scaffolds that have a similar molecular structure natural ECMs [26].

### *4.1.2 Mimicked Self-Organization of the Assembled Collagen*

In natural ECMs, proteins and polysaccharide self-organize into various structures in different tissues. Hence, self-organization is an important bioprocess which is often

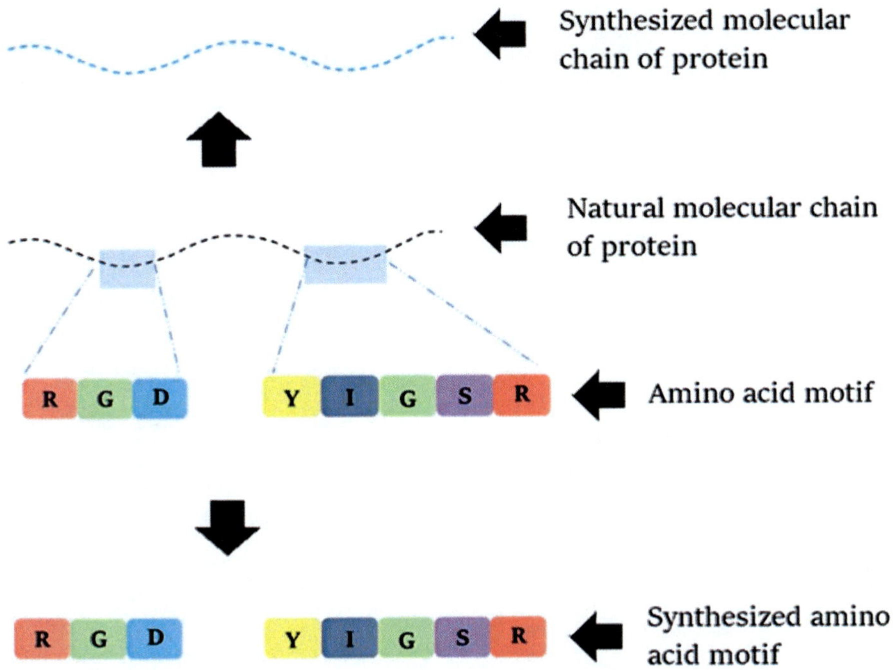


**Fig. 4.1** Scenario of mimicked amino acid motifs and the molecular chain of proteins

used as model to mimic molecular structures in scaffolds. Assembly of collagen in the ECM is often used as a model to mimic self-organization in scaffolds. The structural formation of collagen in different scales is shown in Fig. 4.2.

The structural formation of collagen, in different scales, can self-organize, via the process of assembly. Assembly is started by the dispersion of collagen molecules in an acidic solution. The acidic solution is then adjusted to mimic the condition that leads to self-organization of the collagen molecules into a nano-fibril structure [27, 28]. The nano-structure of collagen fibrils is fixed by adding crosslinking agents into the solution [29]. The size and structure of the fibrils are controlled by the pH concentration, ionic strength in solution, and duration of time during assembly [27, 28].

A pH of 7 is best suitable to form regular collagen fibrils: At a lower or higher pH than 7, collagen often forms into globular or irregular structures [27, 28]. Some research fabricated various sizes as well as structures of collagen fibrils by controlling the pH [27, 28]. These various sizes and structures of collagen fibrils are then used as scaffolds to fit different types of tissue.

At a suitable concentration, the collagen molecules can form into a regular fibril structure [30]. Moreover, at a high concentration the collagen assembly can rapidly reach a steady state and form a gelation better than at a low concentration [30].

The ionic strength in the solution is important for either inducing or reducing the assembly of collagen [30–32]. Generally, fibrillation of collagen leads to gelation

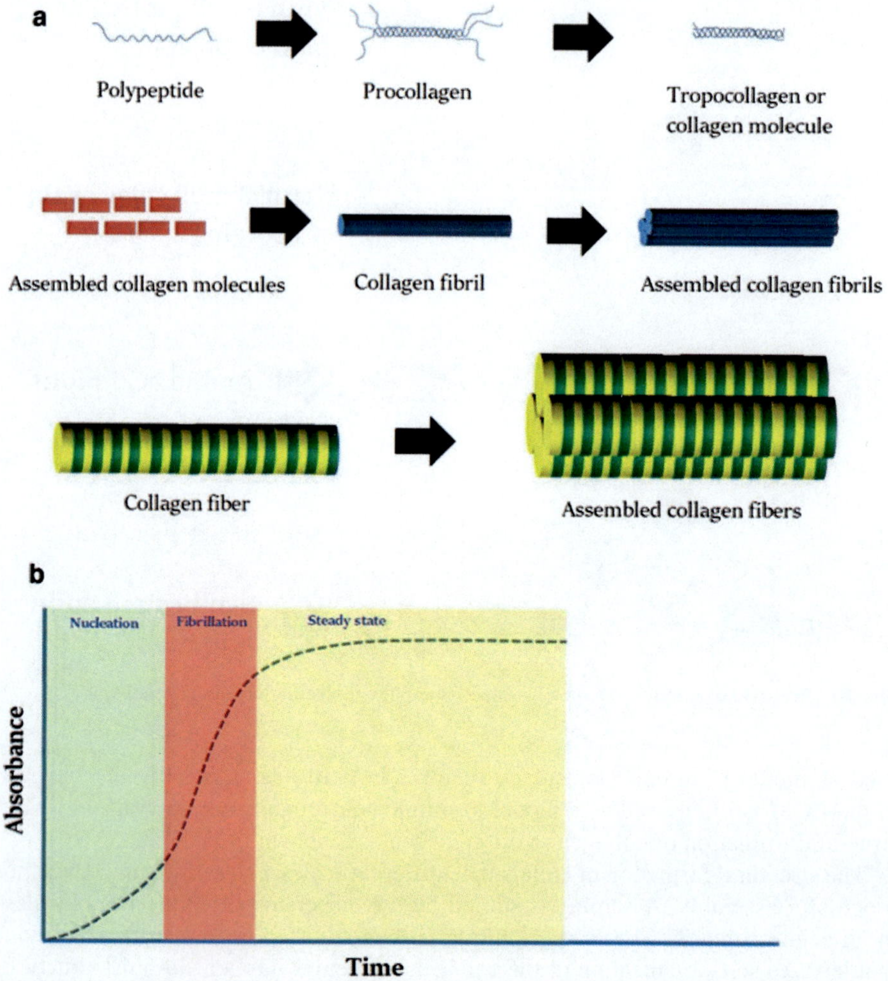


**Fig. 4.2** **a** Self-organization of assembled collagen in different scales and **b** kinetic of assembled collagen

[33]. The performance of the fibril structure can serve to promote cell response that can induce tissue regeneration at the defect site [34]. With a suitable physical as well as mechanical performance, a hydrogel of collagen assembly can support tissue regeneration [35].

### 4.1.3 Mimicked Self-Organization of Assembled Collagen with Mineralization

Biomineralization is the biological process that generates deposition of complex ions on the organic templates [36, 37]. This leads to self-organization of sophisticated structures that perform as natural composite materials. Natural composite materials use the mineralized part as reinforcement and the organic template part as the matrix. For instance, the shells of shrimps, insects, and crabs use the mineralized calcium carbonate as reinforcement and the complex structure of proteins and polysaccharides as the matrix. Such sophisticated structures are guidance for the design and construction of high performance composites for many applications [38–40]. In the case of bone structures for tissue engineering, collagen molecules can self-organize into nano-fibrils as the organic template, which has the pitch for calcium phosphate deposition. Calcium phosphate can self-arrange into a hydroxyapatite structure, and this structure performs as a composite structure that has hydroxyapatite nano-flakes embedded in the collagen nano-fibrils. The composite structure organizes via biomineralization, and the collagen fibrils act as templates for calcium phosphate deposition (Fig. 4.3).

As to the composite structure of collagen nano-fibril/hydroxyapatite nano-flakes, some research has mimicked conditions similar to biomineralization. Collagen molecules were dissolved and dispersed in an acidic solution; this was then mixed with a calcium phosphate solution [41]. Then, the mixed solution was adjusted to a neutral condition before incubation at 37 °C. In this condition, the collagen assembled and biomineralization of the calcium phosphate occurred similar to native ECMs of bone [41]. This mimicked condition can be used to fabricate scaffolds that are similar to the native ECMs of bone [42].

Some research has mimicked the biomineralization of extracellular matrix formation in bone by adding molecules which have effect on formation of calcium phosphate [43–45]. Principally, the molecules self-organize into complex molecules of calcium phosphate before directly approach to collagen fibrils for biomineralization [45]. This approach can be used to mimic the environment of native biomineralization of bone [45]. This approach can also be used to fabricate scaffolds similar to the native structure of bone, and therefore enhancing the biological performance to induce bone regeneration [46]. In addition to the biomineralization of the collagen assembly, other polymers can be used as templates for calcium phosphate deposition.

### 4.1.4 Mimicked Self-Organization of Assembled Collagen with Biological Molecules

Collagen molecules need to assemble with other molecules for the mimicking of ECMs similar to native structures. Examples of these other molecules are laminin,

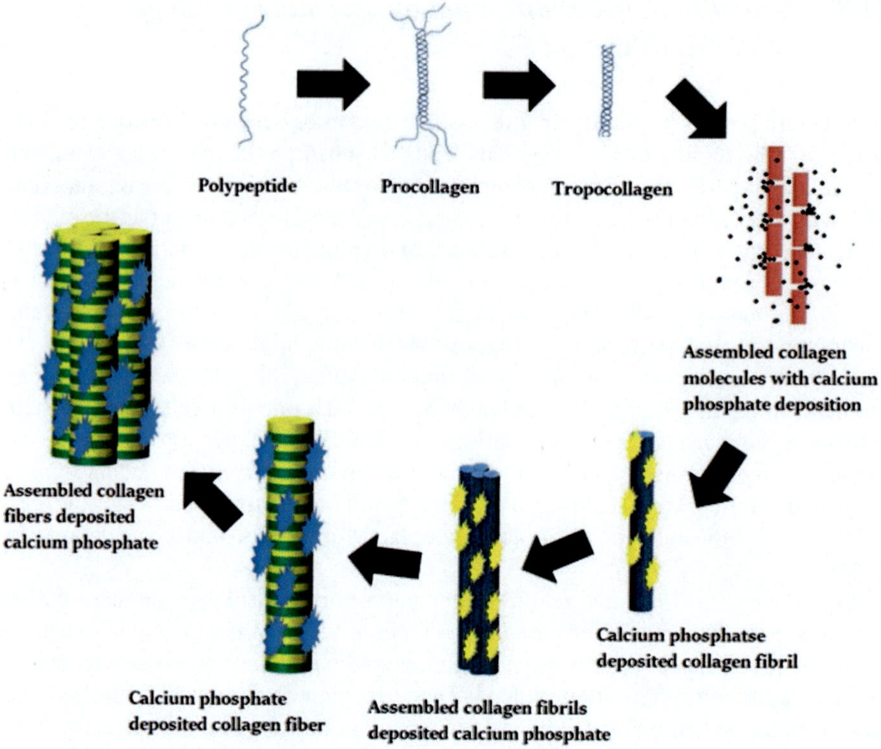


**Fig. 4.3** Biomineralization and structure of a composite structure of collagen nano-fibril/hydroxyapatite nano-plate

fibronectin, and hyaluronic acid [47–49]. These molecules are important components in the ECM. To mimic the structure of the ECM, collagen molecules are mixed with these molecules before adjusting the conditions for suitable self-assembly. The molecules have biological domains that can attach and recognize the collagen molecules. The attachment has the effect on pattern of collagen assembly. The morphology of mimic structure ECMs of assembled collagen incorporated with other components self-organizes into fibrils which are embedded in the matrix [50, 51]. Assembled collagen fibrils are attached during self-organization [52] (Fig. 4.4).

Some research stimulated the structure of scaffolds as native ECMs, by mixing collagen molecules with polymers to mimic the structure of the ECM. For instance, collagen molecules were mixed with some native and derivative polysaccharides that exhibited the same performance as the components in the ECM [39, 53, 54]. The polysaccharides generally had glucose molecules as monomers, for example, methylcellulose, ethyl cellulose, chitosan, and alginate [33, 53, 55, 56]. All of those molecules have a unique functionality of hydrophilicity. When these molecules are mixed with collagen molecules, the assembly of collagen is disturbed, and the kinetic profile of the collagen assembly changes.

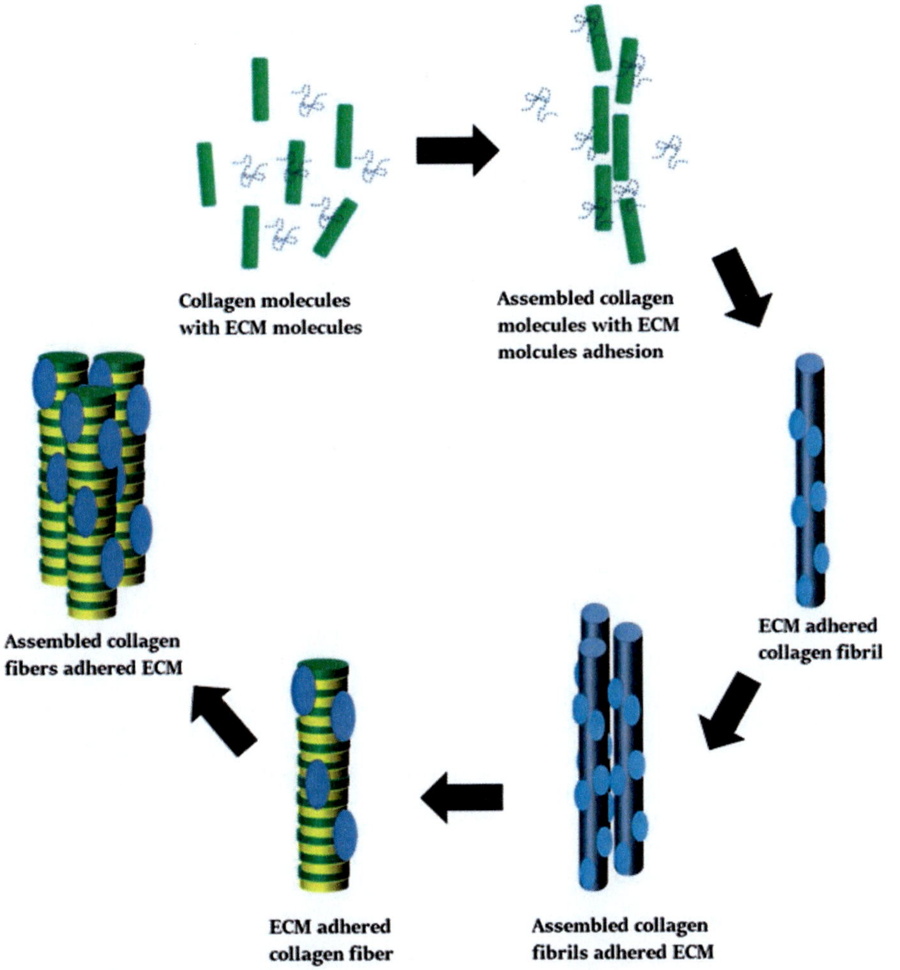


**Fig. 4.4** Collagen fibrils in the matrix with other ECM molecules

Some polysaccharide molecules retard the organization of collagen fibrils by molecular attachment during assembly [56], while in contrast, the crowding effect can induce of collagen assembly [57–59]. Polysaccharide molecules have a charge on their molecules that leads to adherence and molecular attachment onto the collagen molecules. Then, the growth of fibrils, from the nucleating points, is retarded, in effect leading to smaller collagen fibrils [56]. In the case of the crowding effect, the space for collagen assembly is substituted by the crowd of polysaccharide molecules. The residual space has a small volume that can induce collagen assembly [57–60]. The crowding effect can regulate the collagen assembly into a dense fibril structure. Due to the different structures of collagen caused by those effects, some researchers could fabricate scaffolds into different morphologies, so as to fit different types of tissue [61].

## 4.2 Mimicked Self-Organization, Based on Synthetic Molecules

According to the model of collagen assembly in the native environment, some researchers synthesize molecules that can uniquely self-organize into nano-fibril structures. At the nano-scale level, smart molecules have the functionality to self-assemble into a fibril structure similar to native ECMs [62, 63].

Some oligomers or polymers have often been used to construct nano-fibrils as native ECMs. Those oligomers generally have hydrophobic and hydrophilic domains in their structure that can self-assemble into nano-fibrils by hydrophobic interaction, since the core and hydrophilic shell are in contact with the aqueous solution. This can be used to fabricate scaffolds for different kinds of tissue. In the case of bone, the oligomers can mix with saturated calcium phosphate ions, and this mixture organizes into biomineralization for bone formation. The hydrophilic shell surface of the nano-fibrils acts as a template for calcium phosphate deposition (Fig. 4.5) [64, 65].

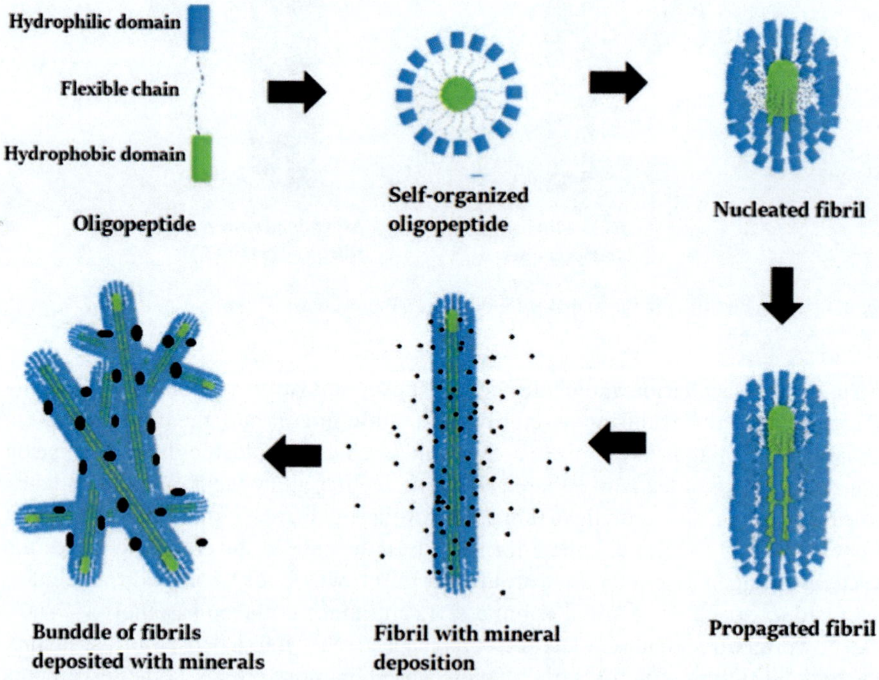


Fig. 4.5 Biomineralization of self-organized molecules

## 4.3 Molecular Modification in Mimicked Molecular Structures

Besides mimicking the environment of native ECM construction, collagen molecules have been modified by grafting amino acids into their molecules. Modified collagen molecules can maintain the assembly functionality [66]. Interestingly, the modified collagen molecules can self-assemble into various structures, depending on molecular components [66–68], so these various structures can then fit different types of tissue [69].

For the modification of scaffolds, polymer is crosslinked or grafted on the molecules by chemical reagents [70, 71]. For modification by crosslinking, polymer was added to reactive molecules. Reactive molecules can generate crosslinking that acts like bridges between the molecules. For grafting, the molecules can react with the chemical groups of silk fibroin, and then, the silk fibroin molecules self-organize into the grafting structure. The molecular structure of the silk fibroin changes following modification by crosslinking and grafting. Said crosslinking leads to self-organization into a network structure, as illustrated in Fig. 4.6.

Crosslinked structures stabilize the molecules; for instance, crosslinked structures have high thermal stability and can provide mechanical strength [72, 73]. The network structure exhibits a swelling behavior in solvents, but does not dissolve [74]. Properties of crosslinked structures depend on their characteristics, length, and density of the crosslinkers [72, 73]. Organization of grafted molecules this is identified in the Fig. 4.7, this structure is a loose structure that has spaces for the diffusion of small molecules, and this diffusion can induce flexibility of the structure leading to increase the water uptake [74].

The molecular modification of polymers has an intense influence on its properties. On the plus side, modification can improve mechanical, physical, chemical, and

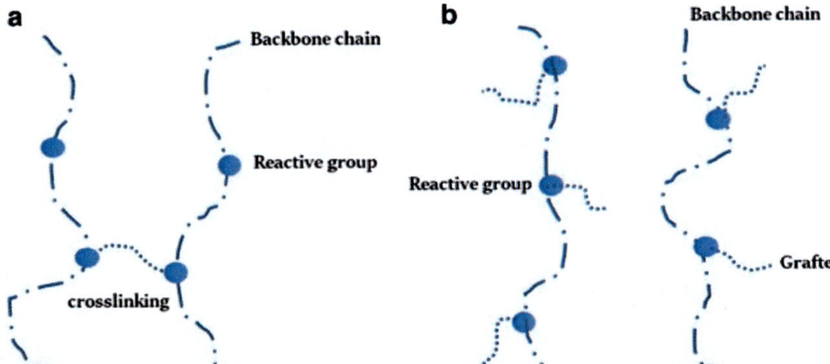


**Fig. 4.6** Molecular structures of crosslinked polymers (**a**) and grafted polymers (**b**)

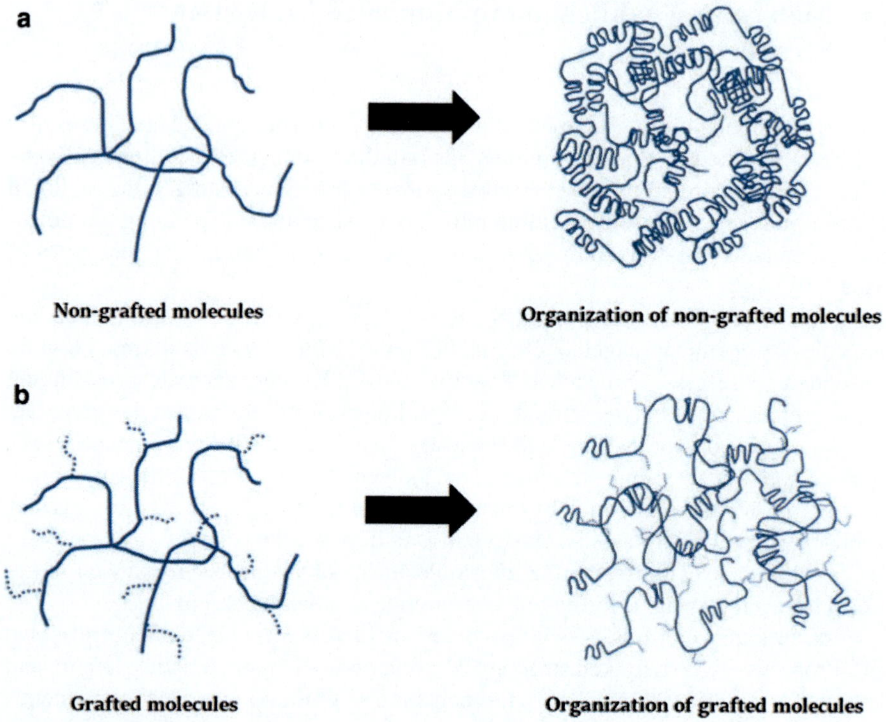


**Fig. 4.7** Organization of non-grafted molecules (**a**) and grafted molecules (**b**)

biological properties [Ref], although, on the negative side, modification can also have a negative influence on the same properties. Choosing a suitable, modified structure that has the properties to fit the application is an important point to consider before application. For instance, in the case of hard tissue, suitable properties are physical and mechanical stability. Therefore, modifying the silk fibroin molecules into a highly regular organization leads to both physical and mechanical stability. Contrary, for soft tissue silk fibroin molecules need a modification that can form a loose structure. A loose structure can provide elasticity, which is more suitable for soft tissue.

## 4.4 Chemical Treatment for Mimicked Molecular Structures

Polymers have been fabricated generally into two- and three-dimensional (2D and 3D) scaffolds. Then, some techniques were used to modify their structure for enhancement of the performance of the scaffolds. For instance, silk fibroin is a protein molecule that can self-organize into regular structures (beta-sheet or alpha-helix) or an amorphous structure (random coils). Silk fibroin has often been treated with some chemical reagents to induce a transformation from an amorphous structure into a regular structure. An example of this is a solution of alcohol, and this has been used to treat silk fibroin scaffolds, as alcohol molecules replace and pull out the residual water molecules, which act as a bridge between the silk fibroin molecules [75]. During stated transformation, the silk fibroin molecules increasingly self-organize into a regular structure, via molecular interaction of the chemical groups [76]. Thus, the structure of the silk fibroin can pack more densely.

In some cases, chemical reagents can diffuse and penetrate into the silk fibroin molecules promoting plasticization [77, 78]. Plasticization can segregate the molecules, and this reduces the molecular interaction of silk fibroin molecules. Importantly, molecular interaction affects the properties of the scaffolds. Strong molecular interaction can induce mechanical strength, physical stability, and barrier properties [79]. Other polymers, besides silk fibroin, can generate molecular transformation during preparation [80]. Said transformations can be applied to mimic the structure of native extracellular matrix, particularly the fibrillation of those polymers.

Detailed in this chapter is the mimicking of molecular structures, as a guidance to create performance scaffolds, which promote cell behavior for enhancement of tissue formation. Nevertheless, to develop the potential of scaffolds, mimicking of morphology and geometry similar to natural tissue is undertaken. The mimicked morphology and geometry in scaffolds will be explained within the next chapter.

## References

1. Negrini, N.C., Bonnetiera, M., Giatsidis, G., Orgill, D.P., Farè, S., Marelli, B.: Tissue-mimicking gelatin scaffolds by alginate sacrificial templates for adipose tissue engineering. Acta Biomat. **87**, 61–75 (2019)
2. Jaipaew, J., Wangkulangkul, P., Meesane, J., Raungrut, P., Puttawibul, P.: Mimicked cartilage scaffolds of silk fibroin/hyaluronic acid with stem cells for osteoarthritis surgery: morphological, mechanical, and physical clues. Mat. Sci. Eng. C-Mater. **64**, 173–182 (2016)
3. Wangkulangkul, P., Jaipaew, J., Puttawibul, P., Meesane, J.: Constructed silk fibroin scaffolds to mimic adipose tissue as engineered implantation materials in post-subcutaneous tumor removal. Mater. Design **106**, 428–435 (2016)
4. Bektas, C.K., Kimiz, I., Sendemir, A., Hasirci, V., Hasirci, N.: A bilayer scaffold prepared from collagen and carboxymethyl cellulose for skin tissue engineering applications. J. Biomat. Sci-Polym. E **29**, 1764–1784 (2018)
5. He, J., Li, D., Liu, Y., Yao, B., Zhan, H., Lian, Q., Lu, B., Lv, Y.: Preparation of chitosan–gelatin hybrid scaffolds with well-organized microstructures for hepatic tissue engineering. Acta Biomater. **5**, 453–461 (2009)
6. Cheema, U., Rong, Z., Kirresh, O., MacRobert, A.J., Vadgama, P., Brown, R.A.: Oxygen diffusion through collagen scaffolds at defined densities: implications for cell survival in tissue models. J. Tissue Eng. Regen. M **6**, 77–84 (2012)
7. Barber, J.G., Handorf, A.M., Allee, T.J., Li, W.J.: Braided nanofibrous scaffold for tendon and ligament tissue engineering. Tissue Eng. Pt A **19**, 1265–1274 (2013)
8. Mandal, B.B., Park, S.H., Gil, E.S., Kaplan, D.L.: Multilayered silk scaffolds for meniscus tissue engineering. Biomat. **32**, 639–651 (2011)

9. Suvarnapathaki, S., Wu, X., Lantigua, D., Nguyen, M.A., Camci-Unal, G.: Hydroxyapatite-incorporated composite gels improve mechanical properties and bioactivity of bone scaffolds. Macromol. Biosci. **20**, 2000176 (2020)
10. Vaquette, C., Kahn, C., Frochot, C., Nouvel, C., Six, J.L., De Isla, N., Luo, L.H., Cooper-White, J., Rahouadj, R., Wang, X.: Aligned poly(L-lactic-co-e-caprolactone) electrospun microfibers and knitted structure: a novel composite scaffold for ligament tissue engineering. J. Biomed. Mater. Res. A **94A**, 1270–1282 (2010)
11. Baek, J., Sovani, S., Choi, W., Jin, S., Grogan, S.P., D'Lima, D.D.: Meniscal tissue engineering using aligned collagen fibrous scaffolds: comparison of different human cell sources. Tissue Eng. Pt. A **24**, 81–93 (2018)
12. Stocco, T.D., Rodrigues, B.V.M., Marciano, F.R., Lobo, A.O.: Design of a novel electrospinning setup for the fabrication of biomimetic scaffolds for meniscus tissue engineering applications. Mater. Lett. **196**, 221–224 (2017)
13. Bansal, S., Mandalapu, S., Aeppli, C., Qua, F., Szczesny, S.E., Mauck, R.L., Zgonis, M.H.: Mechanical function near defects in an aligned nanofiber composite is preserved by inclusion of disorganized layers: Insight into meniscus structure and function. Acta Biomater. **56**, 102–109 (2017)
14. Yu, L., Wei, M.: Biomineralization of collagen-based materials for hard tissue repair. Int. J. Mol. Sci. **22**(2), 944 (2021)
15. Souza, S.E.D., Ginsberg, M.H., Plow, E.F.: Arginyl-glycyl-aspartic acid (RGD): a cell adhesion motif. Trends Biochem. Sci. **16**, 246–250 (1991)
16. Hwang, S., Gim, Y., Gyun, D., Yeon, K., Kim, K., Cha, H.J.: Recombinant mussel adhesive protein Mgfp-5 as cell adhesion biomaterial. J. Biotechnol. **127**, 727–735 (2007)
17. Ruggiero, F., Hägg, P., Pihlajaniemi, T., Huhtala, P.: Recombinant human collagen XV regulates cell adhesion and migration. J. Biol. Chem. **285**, 5258–5265 (2010)
18. Jung, M.Y., Thapa, N., Kim, J.E., Yang, J.D., Cho, B.C., Kim, I.S.: Recombinant tetra-cell adhesion motifs supports adhesion, migration and proliferation of keratinocytes/fibroblasts, and promotes wound healing. Exp. Mol. Med. **39**, 663–672 (2007)
19. Włodarczyk-Biegun, M.K., Werten, M.W.T., Posadowska, U., Storm, I.M., de Wolf, F.A., van den Beucken, J.J.J.P., Leeuwenburgh, S.C.G., Martien, A., Stuart, M.A.C., Kamperman, M.: Nanofibrillar hydrogel scaffolds from recombinant protein-based polymers with integrin- and proteoglycan-binding domains. J. Biomed. Mater. Res. A **104**, 3082–3092 (2016)
20. Li, J., Ding, M., Fu, Q., Tan, H., Xie, X., Zhong, Y.: A novel strategy to graft RGD peptide on biomaterials surfaces for endothelization of small-diamater vascular grafts and tissue engineering blood vessel. J. Mater. Sci: Mater. Med. **19**, 2595–2603 (2008)
21. Zhang, P., Wu, H., Wu, H., Lù, Z., Deng, C., Hong, Z., Jing, X., Chen, X.: RGD-conjugated copolymer incorporated into composite of poly(lactide-co-glycotide) and poly(l-lactide)-grafted nanohydroxyapatite for bone tissue engineering. Biomacromol. **12**(7), 2667–2680 (2011)
22. Ko, E., Yang, K., Shin, J., Cho, S.W.: Polydopamine-assisted osteoinductive peptide immobilization of polymer scaffolds for enhanced bone regeneration by human adipose-derived stem cells. Biomacromol. **14**(9), 3202–3213 (2013)
23. Schmidt, D.R., Kao, W.J.: Monocyte activation in response to polyethylene glycol hydrogels grafted with RGD and PHSRN separated by interpositional spacers of various lengths. J. Biomed. Mater. Res. A **83**, 617–625 (2007)
24. Kim, T.G., Park, T.G.: Biomimicking extracellular matrix: cell adhesive RGD peptide modified electrospun poly(D, L-lactic-co-glycolic acid) nanofiber mesh. Tissue Eng. **12**, 221–233 (2006)
25. Andukuri, A., Kushwaha, M., Tambralli, A., Anderson, J.M., Dean, D.R., Berry, J.L., Sohn, Y.D., Yoon, Y.S., Brotte, B.C., Jun, H.W.: A hybrid biomimetic nanomatrix composed of electrospun polycaprolactone and bioactive peptide amphiphiles for cardiovascular implants. Acta Biomater. **7**, 225–233 (2011)
26. Fittkau, M.H., Zilla, P., Bezuidenhout, D., Lutolf, M.P., Human, P., Hubbell, J.A., Davies, N.: The selective modulation of endothelial cell mobility on RGD peptide containing surfaces by YIGSR peptides. Biomat. **26**, 167–174 (2005)

27. Zhu, S., Yuan, Q., Yin, T., You, J., Gu, Z., Xiong, S., Hu, Y.: Self-assembly of collagen-based biomaterials: preparation, characterizations and biomedical applications. J. Mater. Chem. B **6**, 2650–2676 (2018)
28. Jiang, F., Hörber, H., Howard, J., Müller, D.J.: Assembly of collagen into microribbons: effects of pH and electrolytes. J. Struct. Biol. **148**, 268–278 (2004)
29. Cornwell, K.G., Lei, P., Andreadis, S.T., Pins, G.D.: Crosslinking of discrete self-assembled collagen threads: Effects on mechanical strength and cell–matrix interactions. J. Biomed. Mater. Res. A **80A**, 362–371 (2007)
30. Yan, M., Li, B., Zhao, X., Qin, S.: Effect of concentration, pH and ionic strength on the kinetic self-assembly of acid-soluble collagen from walleye pollock (Theragra chalcogramma) skin. Food Hydrocolloid **29**, 199–204 (2012)
31. Achilli, M., Mantovani, D.: Tailoring mechanical properties of collagen-based scaffolds for vascular tissue engineering: the effects of pH, temperature and ionic strength on gelation. Polymer. **2**(4), 664–680 (2010)
32. Shen, L., Bu, H., Yang, H., Liu, W., Li, G.: Investigation on the behavior of collagen self-assembly in vitro via adding sodium silicate. Int. J. Biol. Macromol. **115**, 635–642 (2018)
33. Yan, M., Jiang, X., Wang, G., Wang, A., Wang, X., Wang, X., Zhao, X., Xu, H., An, X., Li, Y.: Preparation of self-assembled collagen fibrillar gel from tilapia skin and its formation in presence of acidic polysaccharides. Carbohyd. Polym. **233**, 115831 (2020)
34. Moroi, S., Miura, T., Tamura, T., Zhang, X., Ura, K., Takagi, Y.: Self-assembled collagen fibrils from the swim bladder of Bester sturgeon enable alignment of MC3T3-E1 cells and enhance osteogenic differentiation. Mat. Sci. Eng. C-Mater. **94**, 109925 (2019)
35. Patel, B., Xu, Z., Pinnock, C.B., Kabbani, L.S., Lam, M.T.: Self-assembled Collagen-Fibrin Hydrogel Reinforces Tissue Engineered Adventitia Vessels Seeded with Human Fibroblasts. Sci. Rep. **8**, 3294 (2018)
36. Ehrlich, H.: Chitin and collagen as universal and alternative templates in biomineralization. Int. Geol. Rev. **52**, 661–699 (2010)
37. Yao, S., Jin, B., Liu, Z., Shao, C., Zhao, R., Wang, X., Tang, R.: Biomineralization: from material tactics to biological strategy. Adv. Mater. **29**, 1605903 (2017)
38. Torres-Rendon, J.G., Leal-Egaña, A., Walther, A., Schlaad, H., Helmut Cölfen, H., Scheibel, T.R.: Biomineralization of engineered spider silk protein-based composite materials for bone tissue engineering. Mat. **9**(7), 560 (2016)
39. Evans, J.S.: Composite materials design: biomineralization proteins and the guided assembly and organization of biomineral nanoparticles. Mat. **12**(4), 581 (2019)
40. Elsharkawy, S., Mata, A.: Hierarchical biomineralization: from nature's designs to synthetic materials for regenerative medicine and dentistry. Adv. Health. Mater. **7**, 1800178 (2018)
41. Bradt, J.H., Mertig, M., Teresiak, A., Wolfgang, P.W.: Biomimetic mineralization of collagen by combined fibril assembly and calcium phosphate formation. Chem. Mater. **11**(10), 2694–2701 (1999)
42. Jitphuthi, P., Tangtrakulwanich, B., Meesane, J.: Hierarchical porous formation, collagen and mineralized collagen modification of polylactic acid to design mimicked scaffolds for maxillofacial bone surgery. Mater. Today Commun. **13**, 46–52 (2017)
43. Flade, K., Lau, C., Mertig, M., Wolfgang, P.W.: Osteocalcin-controlled dissolution–reprecipitation of calcium phosphate under biomimetic conditions. Chem. Mater. **13**(10), 3596–3602 (2001)
44. Pompe, W., Worch, H., Habraken, W.J.E.M., Simon, P., Kniep, R., Ehrlich, H., Peter, P.: Octa-calcium phosphate – a metastable mineral phase controls the evolution of scaffold forming proteins. J. Mater. Chem. B **3**, 5318–5329 (2015)
45. Simon, P., Grüner, D., Worch, H., Pompe, W., Lichte, H., Khassawna, T.E., Heiss, C., Wenisch, S., Kniep, R.: First evidence of octacalcium phosphate@osteocalcin nanocomplex as skeletal bone component directing collagen triple–helix nanofibril mineralization. Sci. Rep. **8**, 13696 (2018)
46. Liu, Y., Luo, D., Wang, T.: Hierarchical structures of bone and bioinspired bone tissue engineering. Small **12**, 4611–4632 (2016)

47. Chung, E.J., Jakus, A.E., Shah, R.N.: In situ forming collagen-hyaluronic acid membrane structures: mechanism of self-assembly and applications in regenerative medicine. Acta Biomater. **9**(2), 5153–5161 (2013)
48. Sevilla, C.A., Dalecki, D., Hocking, D.C.: Extracellular matrix fibronectin stimulates the self-assembly of microtissues on native collagen gels. Tissue Eng. Part A **16**(12), 3805–3819 (2010)
49. Jain, R., Roy, S.: Designing a bioactive scaffold from coassembled collagen–laminin short peptide hydrogels for controlling cell behavior. RSC Adv. **9**, 38745–38759 (2019)
50. Zhang, L., Li, K., Xiao, W., Zheng, L., Xiao, Y., Fan, H., Zhang, X.: Preparation of collagen–chondroitin sulfate–hyaluronic acid hybrid hydrogel scaffolds and cell compatibility in vitro. Carbohyd Polym. **84**, 118–125 (2011)
51. Schwab, A., Hélary, C., Richards, R.G., Alinia, M., Eglin, D., D'Este, M.: Tissue mimetic hyaluronan bioink containing collagen fibers with controlled orientation modulating cell migration and alignment. Mater. Today Bio. **7**, 100058 (2020)
52. Kadler, K., Hill, A., Canty-Laird, E.G.: Collagen fibrillogenesis: fibronectin, integrins, and minor collagens as organizers and nucleators. Curr. Opin. Cell Biol. **20**, 495–501 (2008)
53. Zulkifli, F.H., Hussain, F.S.J., Rasad, M.S.B.A., Yusoff, M.M.: Improved cellular response of chemically crosslinked collagen incorporated hydroxyethyl cellulose/poly(vinyl) alcohol nanofibers scaffold. J. Biomater. Appl. **29**, 1014–1027 (2014)
54. Rudisill, S.G., DiVito, M.D., Hubel, A., Steinn, A.: In vitro collagen fibril alignment via incorporation of nanocrystalline cellulose. Acta Biomater. **12**, 122–128 (2015)
55. Puttawibul, P., Benjakul, S., Meesane, J.: An in situ hydrogel of mimicked self-assembly type I collagen from shark skin (brownbanded bamboo shark, Chiloscyllium punctatum)/methylcellulose for central nerve system regeneration: preparation and characterization. J. Biomim, Biomater. Biomed. Eng. **14**, 14–29 (2015)
56. Tsai, S.W., Liu, R.L., Hsu, F.Y., Chen, C.C.: A study of the influence of polysaccharides on collagen self-assembly: Nanostructure and kinetics. Biopolymer. **83**, 381–388 (2006)
57. Graham, J., Raghunath, M., Vogel, V.: Fibrillar fibronectin plays a key role as nucleator of collagen I polymerization during macromolecular crowding-enhanced matrix assembly. Biomat. Sci. **7**, 4519–4535 (2019)
58. Zhou, H.X., Rivas, G., Minton, A.P.: Macromolecular crowding and confinement: biochemical, biophysical, and potential physiological consequences. Ann. Rev. Biophys. **37**, 375–397 (2008)
59. Saeidi, N., Karmelek, K.N., Paten, J.A., Zareian, R., DiMasi, E., Ruberti, J.W.: Molecular crowding of collagen: a pathway to produce highly-organized collagenous structures. Biomat. **33**(30), 7366–7374 (2012)
60. De Pieria, A., Rana, S., Korntner, S., Zeugolis, D.I.: Seaweed polysaccharides as macromolecular crowding agents. Int. J. Biol. Macromol. **164**, 434–446 (2020)
61. Chen, C., Loe, F., Blocki, A., Peng, Y., Raghunath, M.: Applying macromolecular crowding to enhance extracellular matrix deposition and its remodeling in vitro for tissue engineering and cell-based therapies. Adv. Drug. Deliver Rev. **63**, 277–290 (2011)
62. Zhang, W., Yu, X., Li, Y., Su, Z., Jandt, K.D., Wei, G.: Protein-mimetic peptide nanofibers: Motif design, self-assembly synthesis, and sequence-specific biomedical applications. Prog. Polym. Sci. **80**, 94–124 (2018)
63. Panda, J.J., Singh, V.: ChauhanShort peptide based self-assembled nanostructures: implications in drug delivery and tissue engineering. Polym. Chem. **5**, 4418–4436 (2014)
64. Wang, L., Guan, X., Yin, H., Moradian-Oldak, J., Nancollas, G.H.: Mimicking the self-organized microstructure of tooth enamel. J. Phys. Chem. C **112**(15), 5892–5899 (2008)
65. Palmer, L.C., Newcomb, C.J., Kaltz, S.R., Spoerke, E.D., Stupp, S.I.: Biomimetic systems for hydroxyapatite mineralization inspired by bone and enamel. Chem. Rev. **108**(11), 4754–4783 (2008)
66. Zhang, J., Tu, X., Wang, W., Nan, J., Wei, B., Xu, C., He, L., Xu, Y., Li, S., Wang, H.: Insight into the role of grafting density in the self-assembly of acrylic acid-grafted-collagen. Inter. J. Biol. Macro **128**, 885–892 (2019)
67. Jiang, T., Xu, C., Liu, Y., Liu, Z., Wall, J.S., Zuo, X., Lian, T., Salaita, K., Ni, C., Pochan, D., Conticello, V.P.: Structurally defined nanoscale sheets from self-assembly of collagen-mimetic peptides. J. Am. Chem. Soc. **136**(11), 4300–4308 (2014)

68. Luo, J., Tong, Y.W.: Self-assembly of collagen-mimetic peptide amphiphiles into biofunctional nanofiber. ACS Nano. **5**(10), 7739–7747 (2011)
69. Zhang, J., Yang, W., Xie, L., Tu, X., Wang, W., Xu, C., Wang, H., Li, S.: Fibrillogenesis of acrylic acid-grafted-collagen without self-assembly property inspired by the hybrid fibrils of xenogeneic collagen. Inter. J. Biol. Macro **163**, 2127–2133 (2020)
70. Oryan, A., Kamalia, A., Moshiri, A., Baharvand, H., Daemi, H.: Chemical crosslinking of biopolymeric scaffolds: current knowledge and future directions of crosslinked engineered bone scaffolds. Inter. J. Biol. Macro. **107**, 678–688 (2018)
71. Cai, Y., Li, J., Poh, C.K., Tan, H.C., Thian, E.S., Fuh, J.Y.H., Sun, J., Tay, B.Y., Wang, W.: Collagen grafted 3D polycaprolactone scaffolds for enhanced cartilage regeneration. J. Mater. Chem. B **1**, 5971–5976 (2013)
72. Haugh, M.G., Jaasma, M.J., O'Brien, F.J.: The effect of dehydrothermal treatment on the mechanical and structural properties of collagen-GAG scaffolds. J. Biomed. Mater. Res. A **89A**, 363–369 (2009)
73. Martíneza, A., Davidenko, B.N., Cameron, R.E.: Tailoring chitosan/collagen scaffolds for tissue engineering: Effect of composition and different crosslinking agents on scaffold properties. Carbohyd. Polym. **132**, 606–619 (2015)
74. Jayakumar, R., Prabaharan, M., Reis, R.L., Mano, J.F.: Graft copolymerized chitosan—present status and applications. Carbohyd. Polym. **62**, 142–158 (2005)
75. Kweon, H.Y., Park, Y.H.: Structural and conformational changes of regenerated Antheraea pernyi silk fibroin films treated with methanol solution. J. Appl. Polym. Sci. **73**, 2887–2894 (1999)
76. Li, M., Tao, W., Kuga, S., Nishiyama, Y.: Controlling molecular conformation of regenerated wild silk fibroin by aqueous ethanol treatment. Polym. Adv. Technol. **14**, 694–698 (2003)
77. Li, X., Zhang, H., He, L., Chen, Z., Tan, Z., You, R., Wang, D.: Flexible nanofibers-reinforced silk fibroin films plasticized by glycerol. Compos. Part B-Eng. **152**, 305–310 (2018)
78. dos Santos, F.V., Yoshioka, S.A., Branciforti, M.C.: Large-area thin films of silk fibroin prepared by two methods with formic acid as solvent and glycerol as plasticizer. J. Appl. Polym. Sci. **138**, 50759 (2021)
79. Leng-Koh, D., Cheng, Y., Teng, C.P., Khin, Y.W., Loh, X.J., Tee, S.Y., Low, M., Ye, E., Yu, H.D., Zhang, Y.W., Han, M.Y.: Structures, mechanical properties and applications of silk fibroin materials. Prog. Polym. Sci. **46**, 86–110 (2015)
80. Mowery, M.D., Menzel, M., Cai, M., Evans, C.E.: Fabrication of monolayers containing internal molecular scaffolding: effect of substrate preparation. Langmuir **14**(19), 5594–5602 (1998)

# Chapter 5
# Mimicked Morphology and Geography in Scaffolds

A suitable morphological and geographical structure for cell residence is an important point that requires consideration before application of polymer scaffolds in tissue engineering. One, suitable morphology for the cells is a porous structure, which has an effect on cell adhesion, proliferation, and migration. Generally, polymers are fabricated into 2D and 3D scaffolds that have suitable structures and functions for tissue engineering [1, 2]. In the case of 2D scaffolds, polymers need to be constructed with rough surface films or fibrous membranes, so as to induce cell adhesion and proliferation, leading to tissue regeneration [2–4]. In the case of 3D scaffolds, polymers need to be fabricated into porous structures that have a suitable morphology for the cells [5]. A suitable morphology has inter-connective pores that can enhance nutrient diffusion and cell migration [Ref]. 2D and 3D scaffolds are shown in Fig. 5.1.

A suitable morphology of pores should enhance cell migration, the diffusion of nutrients, and biological signals that can induce tissue regeneration. Suitable, porous structures depend on the types of tissue. For instance, bone tissue has a dense porous structure that is suitable for osteoblasts, osteoclasts, and osteocytes, whereas for adipose tissue, the porous structure needs to be a loose structure that can enhance nutrient diffusion [6, 7]. Furthermore, the size of the pores has an effect on cell adhesion and proliferation inside the scaffold. For instance, the morphological mimicking of a scaffold, similar to native ECM [5, 8], is often fabricated into sophisticated structures that have a hierarchical pore structure, particularly in the example of trabecular bone, as shown in Fig. 5.2.

In trabecular bone, the morphology has a network of large pores, which have small pores distributed on the wall surface of the large pores [9]. The surface of the large pores is suitable for adhesion of osteoblasts and osteoclasts. For an example of morphological mimicking, scaffolds were fabricated into a sophisticated structure (Fig. 5.3).

 
J. Meesane, *Mimicked Tissue Engineering Scaffolds for Maxillofacial and Articular Cartilage Surgery*, Engineering Materials,
https://doi.org/10.1007/978-981-19-7830-2_5

**Fig. 5.1** **a** 2D scaffold, **b** 3D scaffold

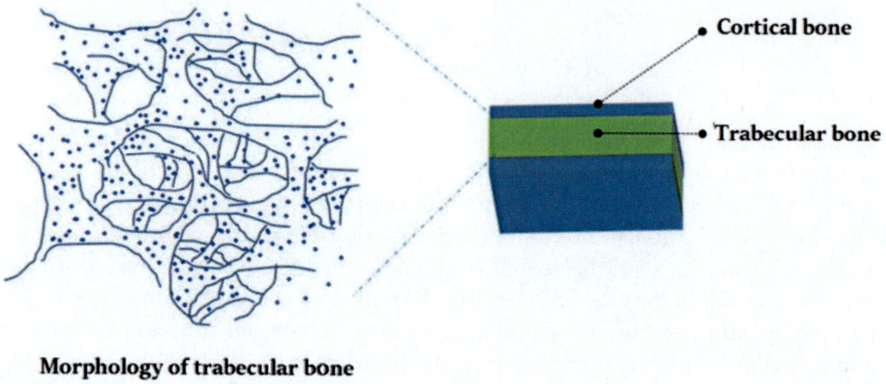


**Fig. 5.2** Morphological structure of trabecular bone

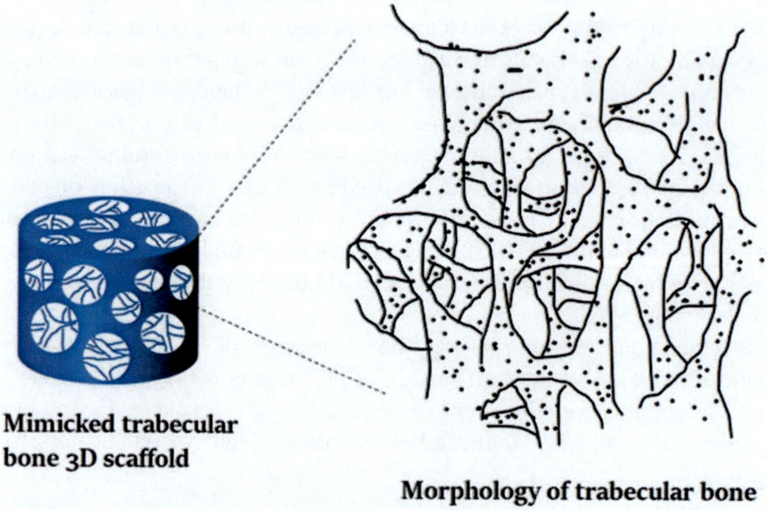


**Fig. 5.3** Mimicked morphology of trabecular bone scaffolds

## 5.1 Fabrication of Mimicked Morphological Structures in Scaffolds

Before mimicking the morphology of scaffolds, suitable polymers are selected for fabrication. The important properties of polymers for scaffolds are classified into: physical, mechanical, and biological functionalities. For physical functionality, polymers must show a physical stability that can maintain the contour shape at the tissue defect site until there is complete, new tissue regeneration [10]. For mechanical functionality, polymers must have the mechanical behavior that fits that of native tissue [11, 12]. In the case of biological functionality, polymers have to induce tissue regeneration and have a biodegradation rate that fits in with new tissue regeneration.

### *5.1.1 Freeze Drying*

Freeze drying is a practical method that has been used often in the fabrication of 3D scaffolds (Fig. 5.4a). In the freeze drying approach the polymer in an aqueous solution, which is then frozen, until the water forms ice crystals in frozen polymer texture [13]. Once this occurs, the ice crystal structure is transformed into a gas state of vapor. The residual structure of these transformed ice crystal structures has a porous structure. The temperature–pressure diagram, during freeze drying, is shown in Fig. 5.4b.

Both pore size and structure of the scaffolds is controlled during the fabrication by two basic parameters: the polymer solution and the process. First, the concentration of the polymer solution has an effect on the pore size and structure [13–15]. In the case of a low concentration, the pore size is larger than that of a high concentration. Furthermore, a thinner wall of the pore is the result of a low concentration. Second, the cooling rate, during the freezing process, is an important parameter in the control of both size and structure of the ice crystals. In the case of a rapid cooling rate, the water forms small ice crystals; this results in thicker walls of the pores; conversely, a slow cooling rate results in thinner walls of the pores.

### *5.1.2 Particle Leaching*

Particle leaching is a practical approach that has been used to fabricate porous scaffolds (Fig. 5.5). To fabricate porous scaffolds, based on particle leaching, particles are added to a polymer melt or solution. Different pore sizes and structures result from the polymer melt or solution and the particles. The concentration of the polymer melt or solution is the main parameter that has an effect on the porous structure. In so saying, a low concentration makes thinner pore walls and porosity than a higher

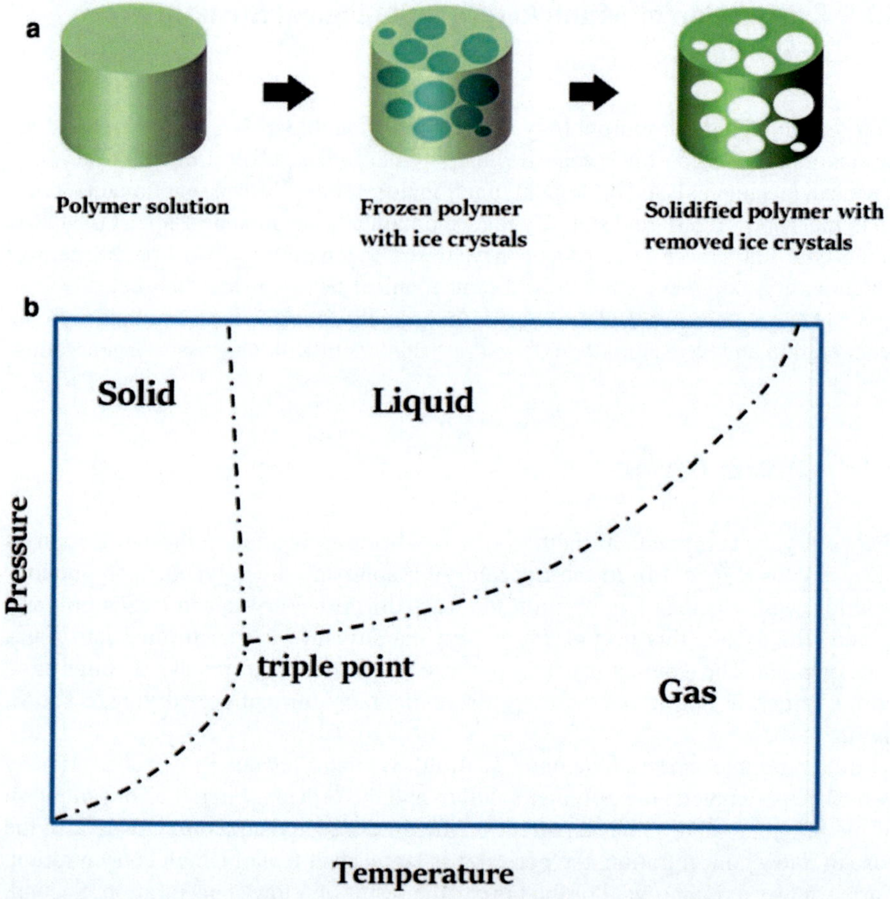


**Fig. 5.4** **a** Freeze drying for scaffold fabrication, **b** Temperature–pressure diagram of freeze drying

concentration [16, 17]. In the case of a very low concentration, the pore walls often show "hole" defects on the surface [16, 17].

There are three main characteristics to be taken into account for the consideration of particles to: (1) particle size, (2) particle shape, and (3) amount of particles [17–19]. First, the size of the pores of a scaffold is proportional to the particle size. Further, if the particles are too large, the pore walls will break, whereas small particles will cause the pore walls to be thick. Second, the particle shape will act as a positive mold to make an imprint in the polymer texture. Therefore, the particle shape has an influence on the structural formation of the porous structure of the scaffold. Third, the amount of particles has an effect on the porous structure and porosity of the scaffold. A low amount of particles will result in a lower porosity and thicker pore walls than a higher amount of particles.

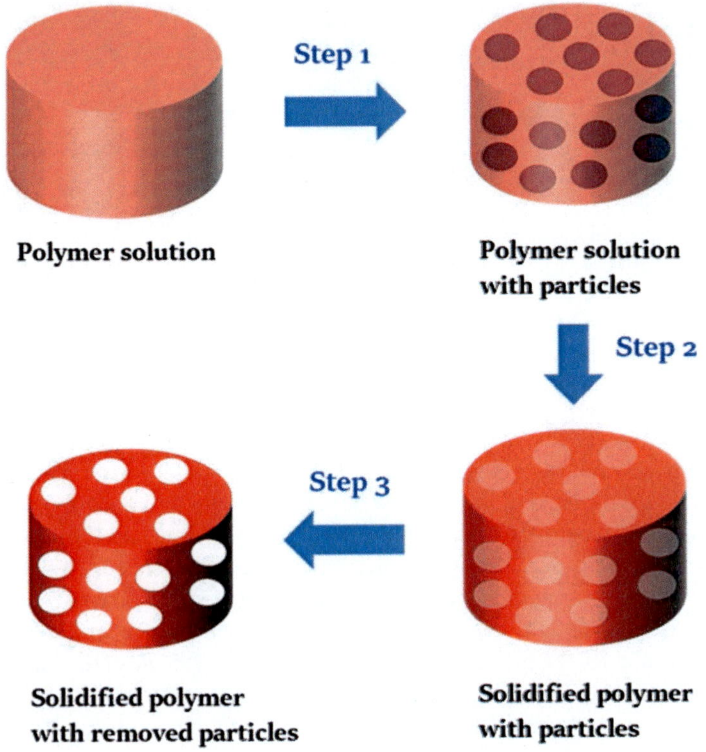


**Fig. 5.5** Particle leaching for scaffold fabrication

Generally, the extracellular matrix (ECM) of native scaffolds in tissue often has an irregular network that has different pore sizes and structures for each type of tissue. For instance, in the case of adipose tissue, the ECM has an irregular, loose network structure that tolerates the growth of adipocytes. This type of structure offers suitable space to support the diffusion of nutrients, growth factors, and cytokines that can induce the formation of new adipose tissue.

### *5.1.3 Particle Leaching with Freeze Drying*

In some cases, combined methods of particle leaching with freeze drying were used in morphological mimicking of hierarchical pores of native bone in the fabrication of scaffolds. The polymers in this technique are fabricated into 3D porous scaffolds by salt leaching, followed by freeze drying.

For fabrication, based on particle leaching, salt particles are generally added to a polymer solution before complete evaporation of the solvent. Then, the salt particles in the samples are leached out by water. This morphological structure has a large

**Fig. 5.6** Particle leaching with freeze drying for scaffold fabrication

network of pores that have some defects as holes on the porous walls [20]. This structure shows a mimicked, loose network structure, much like the morphology of ECM in adipose tissue, particularly in breast tissue. However, this porous structure has insufficient mechanical strength to maintain the contour shapes of the defective tissue. However, the structure can be modified by immersion in a polymer solution, which can seal the defects on the porous walls, before freeze drying [20]. This approach can enhance the mechanical strength to maintain the desired contour shape. When porous scaffolds are immersed in different concentrations of a polymer, the polymer solution can form different structures inside the main pores [20]. First, the polymer molecules in the solution can diffuse and adhere to the surface of the porous walls. The polymer molecules can seal the defect of the wall in the case of a low concentration. On the other hand, in the case of a high concentration, the polymer can fill the main pore and form a porous structure inside the main pores (Fig. 5.6).

### *5.1.4 Three Dimensional (3D) Printing*

3D printing is an approach that has often been used for scaffold fabrications. Principally, 3D printed scaffolds are fabricated from a polymer melt or solution. Establishing a 3D printing is shown in Fig. 5.7.

The first part is the extrusion head or die that has a channel to deliver as well as control the polymer melt or solution from the container. The flow rate of the polymer melt or solution depends on the pressure from the pump. The flow rate of the polymer has an effect on the filament formation of the polymer from the die. A low flow rate will result in slow evaporation of the solvent or slow cooling during solidification of the polymer [21, 22]. This has an influence on the molecular organization of the polymers during solidification. During a slow speed of evaporation and slow cooling rates, the molecules have a long period to self-organize into a greater crystallinity or regularity. In contrast, molecules can form into less crystallinity or regularity

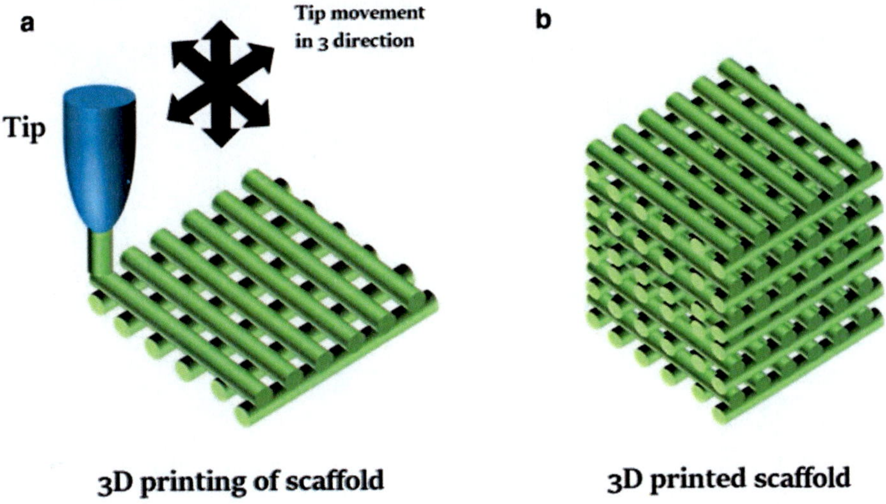


**Fig. 5.7** A) 3D printing of scaffold B) structure of 3D printed scaffold

because they have only a short period to organize themselves during fast evaporation. High crystallinity also has an effect on the shrinkage of the polymer filaments or profile from the die. The organization of polymers, during solidification, results in the physical and mechanical stability of the printed samples [23]. High crystallinity of polymers has high physical stability coupled with high mechanical strength. It is important for the scaffolds to maintain the contour shape at the defect site and also to match the degradation rate of the polymers during tissue regeneration.

The second part is the printer control board that controls the movement of the die. Different movements of the needle have an effect on the constructed form of the samples [24–26]. The samples can be formed into various structures (Fig. 5.8). Different structural formations have different physical and mechanical properties.

The third part is the computer and image software that organizes and constructs the image of the samples [26, 27]. In some cases, images of the defect area are transferred into the computer. The images are organized and constructed into 3D forms by the software. The 3D samples are then printed into a constructed dimensional form by the die. The die is controlled by the printer control board, which is linked with the computer.

The technique of 3D printing has the potential to control the pore size and structure of the scaffolds, which is important for cell adhesion and migration [28]. Cell behavior has an effect on the enhancement of tissue regeneration. Furthermore, 3D printing can construct scaffolds into a suitable shape for filling the tissue defect, which can then be used in the concept of personalized medicine for surgery [27].

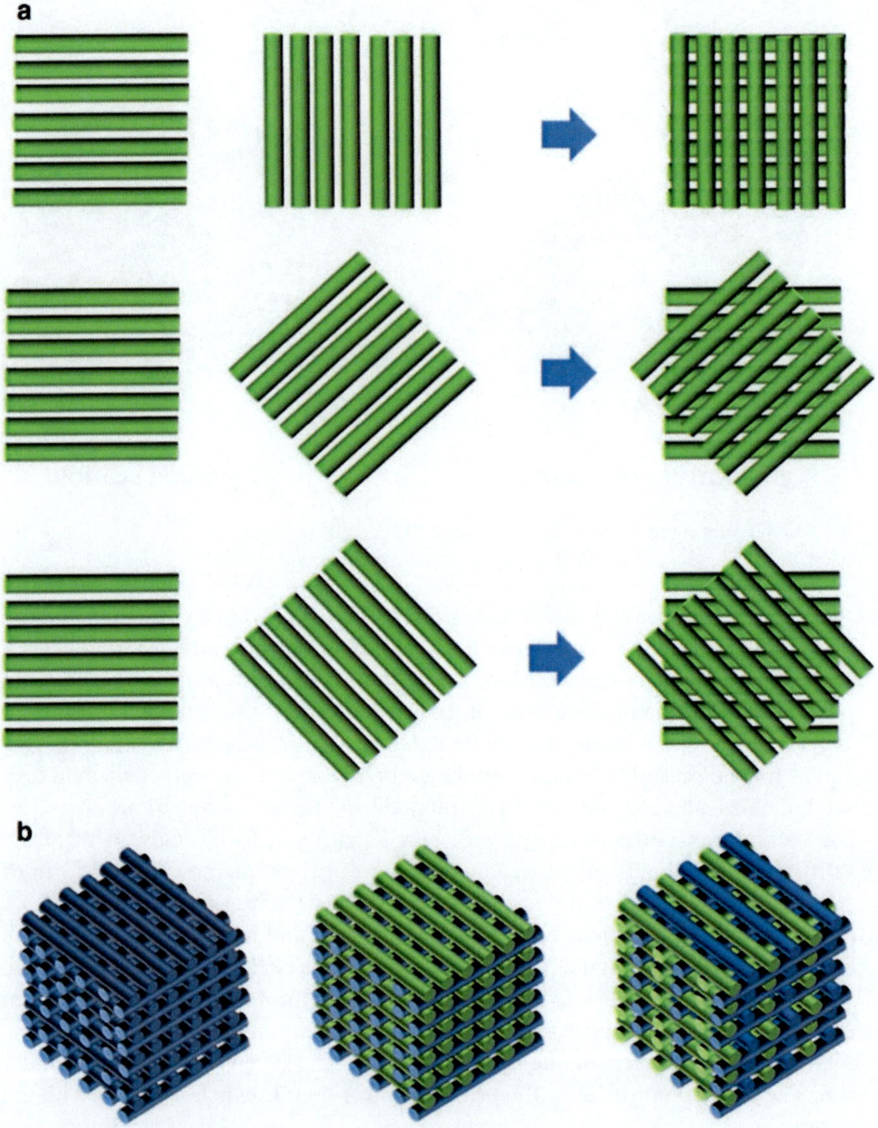


**Fig. 5.8** **a** Different pattern of structural construction of 3D printed scaffolds, **b** Different structural constructed 3D scaffolds

### *5.1.5 Solution Casting*

Solution casting is a practical method for 2D scaffold fabrication (Fig. 5.9). For the fabrication of 2D scaffolds, polymers are dissolved into a solution before being formed into molds. The solvent is then evaporated until those polymers form into solid

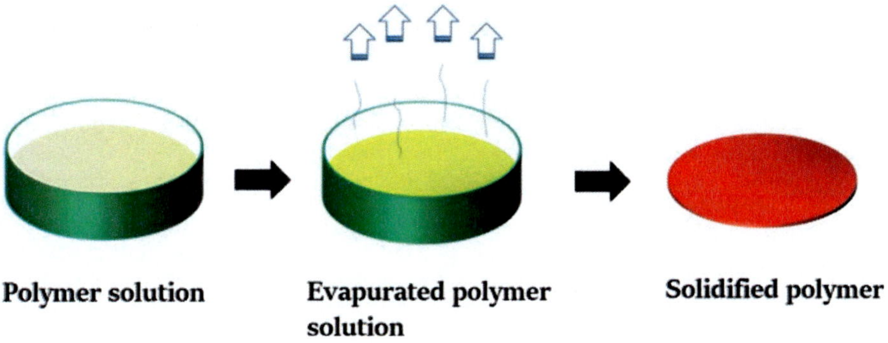


**Fig. 5.9** Solution casting for scaffold fabrication

films. It is important to select polymers that fit the solvent for preparation of a homogeneous solution, before fabrication into the 2D scaffolds or films. When considering suitable solvents for the polymers, solubility parameters of any solvents and polymers must be close to the same values. Generally, topographical and morphological structures differ, and these structural differences depend mainly on two factors: (1) factors from the polymer solution and (2) the process.

The factors of the polymer solution affect the molecular structure of the polymer, which in turn affects the topographical and morphological structures of 2D scaffolds. When considering the regularity of the molecules, the polymers can self-organize into a more orderly crystal structure [29]. In the case of linear chain molecules, the polymer has more flexibility than branched chains [30]. This has an effect on the organization of the molecules into a more crystalline structure of linear chains compared with certain branched chains which demonstrate low flexibility [31]. Furthermore, long chains or high molecular weights have lower mobility than short chains, or low molecular weights [32, 33]. This has an influence on higher crystal structures of low molecular weights compared with high molecular weights [33]. Finally, the concentration of the polymer solution has an effect on the mobility and organization during evaporation. At a low concentration, the polymer has a long time for molecular relaxation that leads to an enhanced higher, regular organization than those at a high concentration [34].

In the case of factors from the process during solidification by evaporation with, have effect on molecular relaxation which results in induction, regular organization [35].

The molecular behavior of the polymers affects the structural organization of the topography along with morphology of solution-casted films. Both, topography and morphology influence cell behaviors on the films, leading to either induction or reduction of tissue regeneration. For instance, in the case of a rough topography, the cells can adhere well onto the surface of the films that are involved with the induction of tissue regeneration in the early state [36]. The highly regular structure of the films exhibits a higher stability to maintain the contour shape of tissue defect during regeneration, than that of a less regular structure. The stability of those films needs optimization of the biodegradation during new tissue regeneration. The films need to maintain the shape during the period of degradation, so as to fit in with new tissue regeneration. Therefore, in the design of a scaffold, it is important to adjust the structural formation, for an optimized performance between stability and biodegradability for each type of tissue.

### *5.1.6 Electro-Spinning*

Electro-spinning is an attractive fabrication method to mimic the structure of a scaffold into 2D forms. Electro-spun scaffolds have a non-woven fiber structure that can mimic from micro-scales to nano-scales. Electro-spinning can also be used to fabricate fibers the size of native collagen in the extracellular matrix [37, 38]. These electro-spun fibers often perform like native collagen that can enhance the cell behaviors of adhesion, proliferation, and migration, and said behaviors lead to the induction of tissue regeneration. Therefore, electro-spinning often has the potential approach to mimic and fabricate high performance scaffolds for tissue engineering.

Principally, mimicked fibrils are formed by an electrical field, and thus, polymer solutions that can generate a charge have been used for fabrication of electro-spun fibers. The outline of the electro-spinning is illustrated in Fig. 5.10. The first part comprises of the syringe, pump, and needle. The syringe serves as a chamber for the polymer solution, while the pump controls the flow rate of the polymer solution in the syringe. Finally, the needle serves as the die to control the shape of the polymer solution, at the tip into a cone shape. The shape is dragged by the electrical field which is the second part of the machine.

This second part is a high voltage generator that controls the voltage of the electrical field. The electrical field is the force to drag the cone shape of the polymer solution into fiber forms. This electrical field has an effect on the structural formation of the fibers, so optimization for a suitable electrical field is important to control the

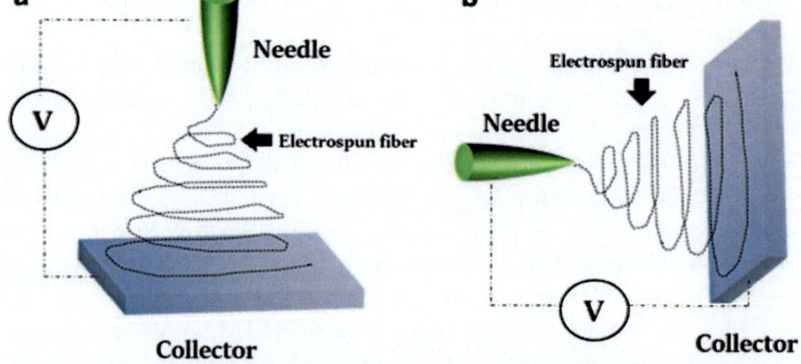


**Fig. 5.10** **a** Outline of electro-spinning in vertical direction, **b** horizontal direction

regularity of the fibers. When the electrical field is not optimized, electro-spun fibers will break or form into beads during the fabrication.

The third part is the collector that forms electro-spin fibers into different geometries. For instance, electro-spun fibers collected by the plate collector have the geometry of membranes, whereby electro-spun fibers from the roll collector have the geometry of a tube. Furthermore, in some cases, the movement of the collector has an effect on the uniaxial alignment of these electro-spun fibers.

Importantly, the alignment of electro-spun fibers has an effect on cell behavior, for instance, adhesion, spreading, proliferation, and migration. Especially, neural cells that are well orientated on alignment substrates can induce the biological functionality of neural cells [39, 40].

Morphological mimicking of scaffolds was constructed into nano- and micro-scale scaffolds. Polymers have been fabricated and considered for molecular organization in the nano-scale. Some scaffolds were fabricated into a membrane that had a non-woven fiber structure; these were in the nano-scale range. Currently, one popular method to mimic nano-fibers of native structure of extracellular matrix is electro-spinning. At the micro-scale level, scaffolds can be fabricated by solution casting, particle leaching, or by freeze drying. In the micro-scale range, morphological formation has been emphasized.

Some scaffolds can combine micro- and nano-construction levels to mimic the structure of the native extracellular matrix. Firstly, the polymers are fabricated into micro-porous scaffolds by either freeze drying or particle leaching. The scaffolds are then covered with nano-fibrils by electro-spinning. In this form of a morphological structure, the nano-fibrils serve as a substrate for cell adhesion that leads to the induction of cell spreading and migration. All approaches to mimic the morphology of the extracellular matrix have the aim of inducing a cell behavior that can enhance tissue regeneration.

## 5.2 Construction of Mimicked Geometrical Structure in Scaffolds

In the case of geometrical mimicking, polymeric scaffolds, that have different morphological structures, have been constructed into different geometrical structures. The geometrical structures have to fit the defect site of the tissue. In most cases, simple geometrical structures have been constructed by elements of 2D and 3D forms, upon which 2D and 3D forms are then constructed into sophisticated, geometrical structures.

### 5.2.1 *Construction of Hydrogel, with Porous Scaffolds, into Biphasic Scaffolds*

Scaffolds have often been combined into two parts for geometrical mimicking of biphasic tissue, such as bone/ligament tissue or bone/cartilage tissue [41, 42]. Viscous hydrogels have been combined with 3D porous scaffolds for bone/cartilage tissue because viscous hydrogels perform as cartilage tissue. 3D porous scaffolds have the morphology and performance for bone tissue (Fig. 5.11).

The cartilage section requires materials that can resist compressive forces during movement. Hydrophilic polymers, that have a high molecular weight, have often been used to construct the cartilage section of biphasic scaffolds. This is due to the hydrophilic and high molecular weight properties of the selected polymer being similar to the extracellular matrix of cartilage. Furthermore, the polymer often demonstrates enhanced cell migration that leads to the induction of tissue regeneration. When considering the morphology; hydrogels require a network structure that has large pores. The pores should have a size similar to the chondrocytes. The selected materials should have a porous structure of the scaffolds that is similar to bone morphology. Furthermore, the morphology of the scaffolds needs to hold the contour shape of the defect, until new bone regeneration is complete.

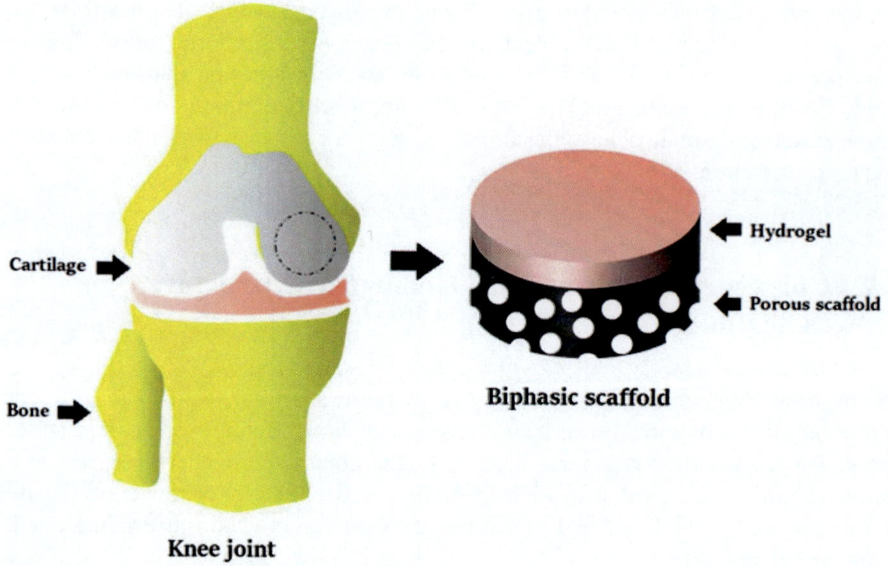


**Fig. 5.11** Bone/cartilage biphasic tissue scaffold

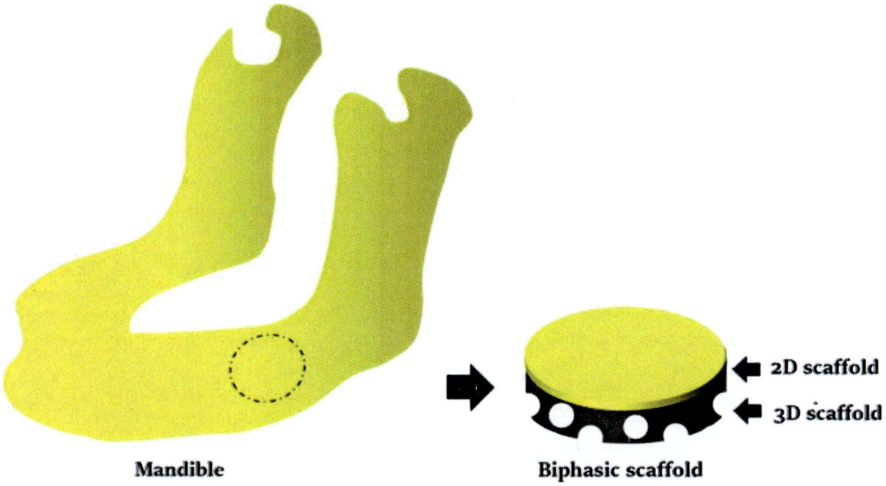


**Fig. 5.12** Bone/mucosa biphasic tissue scaffold

### *5.2.2 Construction of 2D with 3D Scaffolds into Biphasic Scaffolds*

Some areas of maxillofacial tissue have biphasic bone/mucosa tissue. In this case, thin films have been fabricated and combined with 3D porous scaffolds (Fig. 5.12). Film scaffolds induce mucosa tissue regeneration and prevent tissue diffusion into the bone phase. 3D scaffolds induce bone regeneration. In the case of mucosa or other soft tissue, different porous membranes can be combined into biphasic scaffolds. For example, in skin tissue, the first layer of the scaffold can be fabricated into a membrane that is suitable for keratinocytes for the epidermis layer, while the second layer must be fabricated into a membrane that is suitable for fibroblasts for the dermis layer.

The main contents of this chapter discuss the mimicked structure of scaffolds similar to native tissue in different scales. Based on the principle of mimicked structure, we can fabricate performance scaffolds in tissue engineering for different types of defect areas. The potential of scaffolds to enhance functionality will be explained in the next chapter.

## References

1. Zhao, P., Gu, H., Mi, H., Rao, C., Fu, J., Turng, L.S.: Fabrication of scaffolds in tissue engineering: a review. Front. Mech. Eng. **13**, 107–119 (2018)
2. Keirouz, A., Chung, M., Kwon, J., Fortunato, G., Radacsi, N.: 2D and 3D electrospinning technologies for the fabrication of nanofibrous scaffolds for skin tissue engineering: a review. Wires **12**, e1626 (2020)

3. Khorshidi, S., Solouk, A., Mirzadeh, H., Mazinani, S., Lagaron, J.M., Sharifi, S., Ramakrishna, S.: A review of key challenges of electrospun scaffolds for tissue-engineering applications. J. Tissue Eng. Regen M. **10**, 715–738 (2016)
4. Jafari, M., Paknejad, Z., Rad, M.R., Motamedian, S.R., Eghbal, M.J., Nadjmi, N., Khojasteh, A.: Polymeric scaffolds in tissue engineering: a literature review. J. Biomed. Mater. Res. B **105**, 431–459 (2017)
5. Murphy, C.M., Haugh, M.G., Fergal, O'Brien, J.: The effect of mean pore size on cell attachment, proliferation and migration in collagen–glycosaminoglycan scaffolds for bone tissue engineering. Biomaterial **31**, 461–466 (2010)
6. Wu, S., Liu, X., Yeung, K.W.K., Liu, C., Yang, X.: Biomimetic porous scaffolds for bone tissue engineering. Mat. Sci. Eng. R **80**, 1–36 (2014)
7. Laschke, M.W., Kleer, S., Scheuer, C., Schuler, S. Garcia, P., Eglin, D., Alini, M., Menger, M.D.: Vascularisation of porous scaffolds is improved by incorporation of adipose tissue-derived microvascular fragments. Eur. Cells Mater. **24**, 266–277 (2012)
8. Sangkert, S., Kamolmatyakul, S., Gelinsky, M., Meesane, J.: 3D printed scaffolds of alginate/polyvinylalcohol with silk fibroin based on mimicked extracellular matrix for bone tissue engineering in maxillofacial surgery. Mater. Today Commun. **26**, 102140 (2021)
9. Presbítero, G., Gutiérrez, D., Lemus-Martínez, W.R., Vilchez, J.F., García, P., Arizmendi-Morquecho, A.: Assessment of quality in osteoporotic human trabecular bone and its relationship to mechanical properties. Appl. Sci. **11**, 5479 (2021)
10. Zhang, J., Wehrle, E., Vetsch, J.R., Paul, G.R., Rubert, M., Müller, R.: Alginate dependent changes of physical properties in 3D bioprinted cell-laden porous scaffolds affect cell viability and cell morphology. Biomed. Mater. **14**, 065009 (2019)
11. Sabree, I., Gough, J.E., Derby, B.: Mechanical properties of porous ceramic scaffolds: influence of internal dimensions. Ceram. Int. **41**, 8425–8432 (2015)
12. Wang, Y.F., Barrera, C.M., Dauera, E.A., Gua, W., Andreopoulosa, F., Huang, C.Y.C.: Systematic characterization of porosity and mass transport and mechanical properties of porous polyurethane scaffolds. J. Mech. Behav. Biomed. Mater. **65**, 657–664 (2017)
13. O'Brien, F.J., Harley, B.A., Yannas, I.V., Gibson, L.: Influence of freezing rate on pore structure in freeze-dried collagen-GAG scaffolds. Biomaterial **25**, 1077–1086 (2004)
14. Lopes, L., Simone, R., Silva, S., Pedro, R., Alexandra, P., Marques, P., Filipe, J., Tiago, M., Silva, H., Reis, R.L.: Influence of freezing temperature and deacetylation degree on the performance of freeze-dried chitosan scaffolds towards cartilage tissue engineering. Eur. Polym. J. **95**, 232–240 (2017)
15. Hu, Y., Grainger, D.W., Winn, S.R., Hollinger, J.O.: Fabrication of poly($\alpha$-hydroxy acid) foam scaffolds using multiple solvent systems. J. Biomed. Mater. Res. **59**, 563–572 (2002)
16. Zhou, Q., Gong, Y., Gao, C.: Microstructure and mechanical properties of poly(L-lactide) scaffolds fabricated by gelatin particle leaching method. J. Appl. Polym. **98**, 1373–1379 (2005)
17. Prasad, A., Sankar, M.R., Katiyar, V.: State of art on solvent casting particulate leaching method for orthopedic scaffolds fabrication. Mater. Today-Proc. **4**, 898–907 (2017)
18. Reignier, J., Huneault, M.A.: Preparation of interconnected poly($\varepsilon$-caprolactone) porous scaffolds by a combination of polymer and salt particulate leaching. Polymer **47**, 4703–4717 (2006)
19. Yang, Q., Chen, L., Shen, X., Tan, Z.: Preparation of polycaprolactone tissue engineering scaffolds by improved solvent casting/particulate leaching method. J. Macromol. Sci. Phys. **45**, 1171–1181 (2006)
20. Makaya, K., Terada, S., Ohgo, K., Asakura, T.: Comparative study of silk fibroin porous scaffolds derived from salt/water and sucrose/hexafluoroisopropanol in cartilage formation. J. Biosci. Bioeng. **108**, 68–75 (2009)
21. Wangkulangkul, P., Jaipaew, J., Puttawibul, P., Meesane, J.: Constructed silk fibroin scaffolds to mimic adipose tissue as engineered implantation materials in post-subcutaneous tumor removal. Mater. Design **106**, 428–435
22. Northcutt, L.A., Orski, S.V., Migler, K.B., Kotula, A.P.: Effect of processing conditions on crystallization kinetics during materials extrusion additive manufacturing. Polymer **154**, 182–187 (2018)

23. Schiavone, N., Verney, V., Askanian, H.: Effect of 3D printing temperature profile on polymer materials behavior. 3D Printing and Additive Manuf. **7**, 311–325 (2020)
24. Ma, Z., Hong, Y., Nelson, D.M., Pichamuthu, J.E., Leeson, C.E., Wagner, W.R.: Biodegradable polyurethane ureas with variable polyester or polycarbonate soft segments: effects of crystallinity, molecular weight, and composition on mechanical oroperties. Biomacromol. **12**(9), 3265–3274 (2011)
25. Baptista, R., Guedes, M., Pereir, M.F.C., Maurício, A., Carrelo, H., Cidade, T.: On the effect of design and fabrication parameters on mechanical performance of 3D printed PLA scaffolds. Bioprinting **20**, e00096 (2020)
26. Cubo-Mateo, N., Rodríguez-Lorenzo, L.M.: Design of thermoplastic 3D-printed scaffolds for bone tissue engineering: influence of parameters of "hidden" importance in the physical properties of scaffolds. Polymers **12**(7), 1546 (2020)
27. Jung, J.W., Lee, J.S., Cho, D.W.: Computer-aided multiple-head 3D printing system for printing of heterogeneous organ/tissue constructs. Sci. Rep. **6**, 21685 (2016)
28. Yen, H.H., Stathopoulou, P.G.: CAD/CAM and 3D-printing applications for alveolar ridge augmentation. Curr. Oral Health Rep. **5**, 127–132 (2018)
29. Motealleh, A., Dorri, P., Schäfer, A.H., Kehr, N.S.: 3D bioprinting of triphasic nanocomposite hydrogels and scaffolds for cell adhesion and migration. Biofabrication **11**, 035022 (2019)
30. Rungswang, W., Jarumaneeroj, C., Patthamasang, S., Phiriyawirut, P., Jirasukho, P., Soontaranon, S., Rugmai, S., Hsiao, B.S.: Influences of tacticity and molecular weight on crystallization kinetic and crystal morphology under isothermal crystallization: Evidence of tapering in lamellar width. Polymer **172**, 41–51 (2019)
31. Kulagina, T.P., Kurmaz, S.V., Grachev, V.P., Tarasov, V.P.: Topological structure and molecular mobility of linear and branched poly(meth)acrylates studied by NMR relaxation. Russ. Chem. Bull. **60**, 1500–1504 (2011)
32. Zhang, X., Li, Z., Yang, H., Sun, C.C.: Molecular dynamics simulations on crystallization of polyethylene copolymer with precisely controlled branching. Macromolecules **37**(19), 7393–7400 (2004)
33. Acevedo, C.A., Sánchez, E., Díaz-Calderón, P., Blaker, J.J., Enrione, J., Quero, F.: Synergistic effects of crosslinking and chitosan molecular weight on the microstructure, molecular mobility, thermal and sorption properties of porous chitosan/gelatin/hyaluronic acid scaffolds. J. Appl. Polym. Sci. **134**, 44772 (2017)
34. Effect of polymer molecular weight on the crystallization behavior of indomethacin amorphous solid dispersions. Cryst. Growth Des. **17**(6), 3142–3150 (2017)
35. Kos, I., Ivanova, V.A., Chertovich, A.V.: Crystallization of semiflexible polymers in melts and solutions. Soft Matter **17**, 2392–2403 (2021)
36. Udayakumar, M., Kollár, M., Kristály, F., Leskó, M., Szabó, T., Marossy, K., Tasnádi, I., Németh, Z.: Temperature and time dependence of the solvent-induced crystallization of poly(l-lactide). Polymers **12**(5), 1065 (2020)
37. Zareidoost, A., Yousefpour, M., Ghaseme, B., Amir, A.: The relationship of surface roughness and cell response of chemical surface modification of titanium. J. Mater. Sci. Mater. Med. **23**(6), 1479–1488 (2012)
38. Bhowmick, S., Rother, S., Zimmermann, H., Lee, P.S., Moeller, S., Schnabelrauch, M., Koul, V., Jordan, R., Hintze, V., Scharnweber, D.: Biomimetic electrospun scaffolds from main extracellular matrix components for skin tissue engineering application—The role of chondroitin sulfate and sulfated hyaluronan. Mat. Sci. Eng. C-Mater. 79, 15–22 (2017)
39. Watcharajittanont, N., Putson, C., Pripatnanont, P., Meesane, J.: Electrospun polyurethane fibrous membranes of mimicked extracellular matrix for periodontal ligament: Molecular behavior, mechanical properties, morphology, and osseointegration. J. Biomater. Appl. **34**(6), 753–762 (2020)
40. Karimi, A., Karbasi, S., Razavi, S., Zargar, E.N.: Poly(hydroxybutyrate)/chitosan Aligned Electrospun Scaffold as a Novel Substrate for Nerve Tissue Engineering. Adv Biomed Res **7**, 44 (2018)

41. Subramanian, A., Krishnan, U.M., Sethuraman, S.: Fabrication of uniaxially aligned 3D electrospun scaffolds for neural regeneration. Biomed. Mater. **6**(2), 025004 (2011)
42. Liao, J., Tian, T., Shi, S., Xie, X., Ma, Q., Li, G., Lin, Y.: The fabrication of biomimetic biphasic CAN-PAC hydrogel with a seamless interfacial layer applied in osteochondral defect repair. Bone Res. **5**, 17018 (2017)

# Chapter 6
# Mimicked Physical and Mechanical Functions in Scaffolds

Physical and mechanical functions are an important point to consider for the mimicking of scaffolds similar to natural tissue, so mimicked physical and mechanical function, as that of natural tissue, is used in the guidance to create scaffolds. There are many different tissues in the organs of the human body, and each tissue has different physical and mechanical functions, which fit to their organs and systems. For instance, bone tissue has more mechanical strength and stiffness than skin tissue. This is because bone tissue in the skeletal system, which has to resist high mechanical force during the movement of the body. Therefore, to understand the physical and mechanical properties of each tissue is the guidance to mimic performance scaffolds. In this chapter, mimicked physical and mechanical functions of the tissues is explained, and approaches for mimicking physical and mechanical functions in scaffolds are described.

## 6.1 Mimicking of Physical Function in Scaffolds

Generally, scaffolds are often structurally mimicked for their physical and mechanical properties [1]. For physical mimicking, the swelling behavior of scaffolds is often considered, as swelling behavior is important to mimic the wet texture of scaffolds, so as to be similar to the native extracellular matrix. Each type of tissue has a different swelling behavior [1]. In some cases, the tissue needs a high supply of blood and nutrients, as in breast and liver tissue, so scaffolds for these tissues should also have a loose morphological structure, with a high swelling behavior [2–7]. Some scaffolds have been created into structure and morphology that can support a sufficient blood supply and maintain the shape of the tissue [7].

The first approach to mimic physical function is to construct scaffolds into a loose structure that can retain a high water volume that results in a high swelling behavior. A simple technique to construct loose morphological structures consists of freeze drying

 
J. Meesane, *Mimicked Tissue Engineering Scaffolds for Maxillofacial and Articular Cartilage Surgery*, Engineering Materials,
https://doi.org/10.1007/978-981-19-7830-2_6

at a low concentration, in combination with large particle leaching [6]. Scaffolds created from this technique have a morphological structure of large, interconnected pores. In the case of hydrogel, it needs this network of loose structures for enhancement of the swelling behavior (Fig. 6.1a). Furthermore, to promote swelling behavior, based materials with hydrophilic molecules are selected for hydrogel (Fig. 6.1b).

A secondary approach, to mimic the physical function of swelling, is blending the scaffolds with hydrophilic polymers that can induce water uptake and swelling

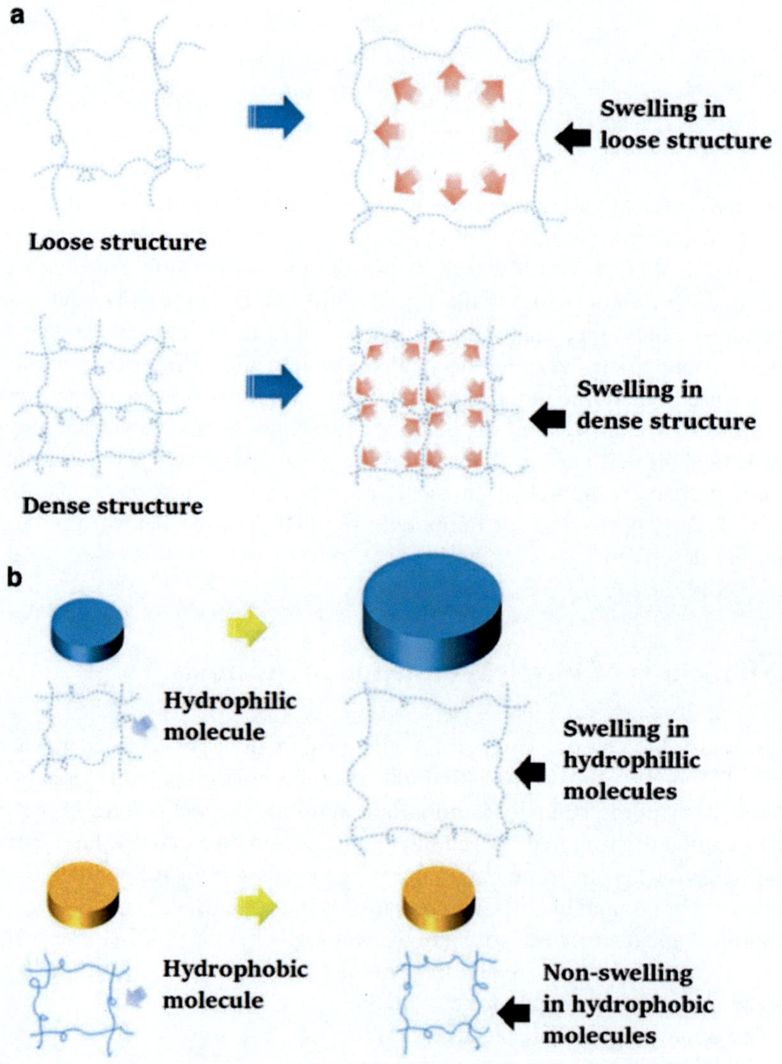


**Fig. 6.1** **a** Loose structure of the network of hydrogel, **b** based materials with hydrophilic molecules for hydrogel

in the same characteristic way as hydrogels [8]. Swelling, due to the uptake of water, enhances nutrient transport, and this transportation of nutrients induces tissue regeneration. Each type of tissue has a different swelling behavior that is dependent on their function. For instance, in the case of cartilage a scaffold is used for replacement at tissue defect. This scaffold shows the dimensions for fitting into the defect site (Fig. 6.2a). Furthermore, it requires high swelling behavior or water uptake, which is similar to natural cartilage tissue (Fig. 6.2b). Scaffolds with high swelling behaviors support nutrient transportation and show resistance of compressive forces at the cartilage (Fig. 6.3c). Therefore, designing suitable scaffolds for cartilage tissue engineering requires a loose structure that has hydrophilic polymers in their structure. Generally, anionic hyaluronic acid can maintain a high water volume and has, therefore, been blended with other polymers as scaffolds in cartilage tissue engineering [9, 10].

In the case of adipose tissue around the breast area, the tissue needs a high amount of nutrients from the blood supply [11, 12]. Furthermore, adipose tissue in the breast area has to maintain its contour shape during tissue regeneration, so as to repair a defect or after tumor removal. Therefore, a loose structure of hydrophilic scaffolds that can enhance the swelling behavior, while maintaining the contour shape, is needed as a substitute for the defect or tumor removal site of the breast [13, 14]. For instance, injectable hydrogel with pre-adipocyte is often filled into the removal area of the breast for enhancement of adipose tissue formation (Fig. 6.3a). Based on the loose structure of said scaffold, it supports cell regulation during adipose tissue formation (Fig. 6.3b).

Another example of this is bone, which requires certain physical characteristics to maintain its shape at the defect site during new tissue formation (Fig. 6.4a). Furthermore, bone tissue also needs a suitable structure for the transport of nutrients from the blood supply. Therefore, scaffolds for bone tissue engineering need hydrophilic polymers in their structure to enhance blood supply. These scaffolds require a low swelling behavior that refers to their physical stability. Normally, a physically, stable scaffold is suitable for cell adhesion. This scaffold has the potential to maintain contour shape at the defect without collapse during tissue formation (Fig. 6.4b). Furthermore, molecules in this scaffold show low extension, which is suitable for cell migration in their structure (Fig. 6.4b). To enhance physical stability, some scaffolds are often blended with calcium phosphate particles, which are able to induce bone tissue formation [15–17]. Other scaffolds are mixed with some bioactive particles, to enhance physical stability and induce bone tissue formation [18–20]. Hence, understanding the physical function is important in designing performance scaffolds to mimic the physical function of scaffolds that are suitable for each type of tissue.

## 6.2 Mimicking of Mechanical Function in Scaffolds

The mechanical function of scaffolds is the clue which has an effect on the different expression of cell behaviors [21, 22]. For flexible or low stiffness scaffolds, cells show

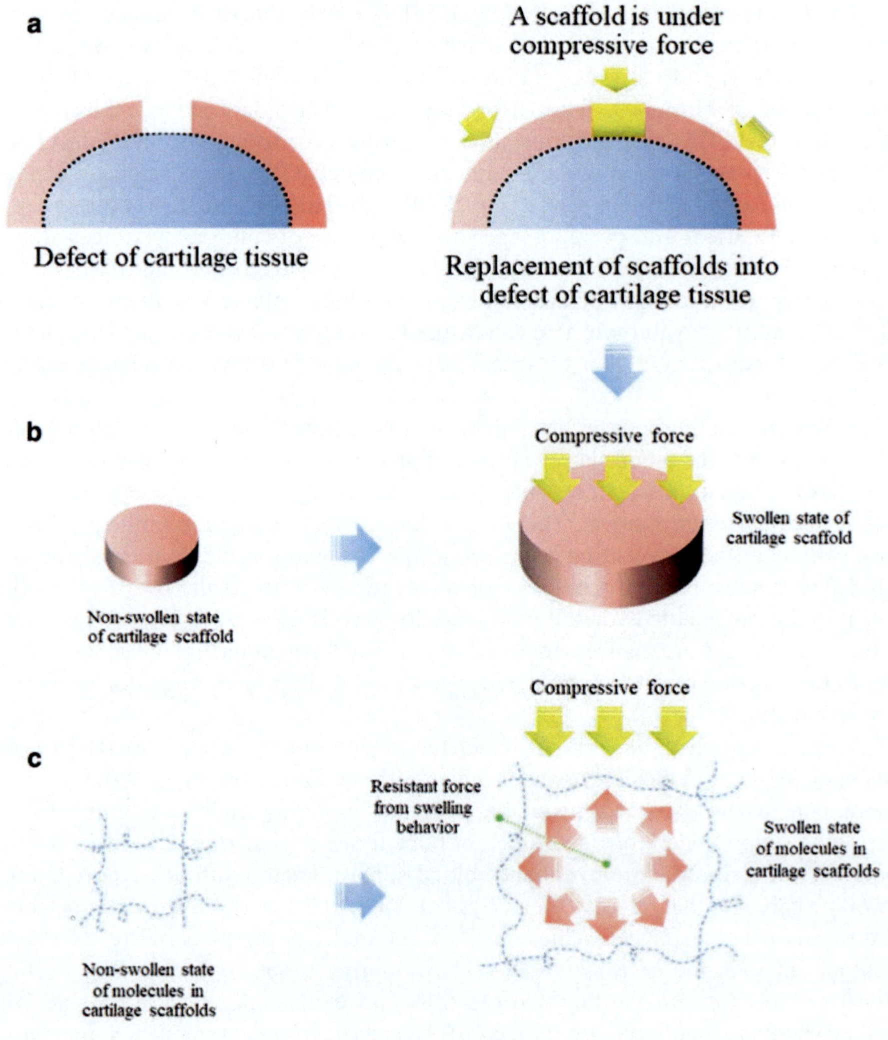


**Fig. 6.2** **a** Scaffold at defect of cartilage, **b** compressive force on the swollen scaffold at the defect of cartilage, and **c** response of swollen scaffold on compressive force at the defect of cartilage

poor adhesion on the surface. This leads to the spherical formation of cells which are generally in the loose structure of extracellular matrix [21–23]. For instance, soft scaffolds are often used for adipose tissue [23–25] (Fig. 6.5); wherein, in the case of stiffness or hard scaffolds, cells show good adhesion, which has an effect on spreading behaviors [23, 26, 27] (Fig. 6.5). This scaffold is suitable for bone tissue which demonstrates more mechanical stiffness than the other tissues. Additionally, these cell behaviors lead to different morphogenesis of tissue [23].

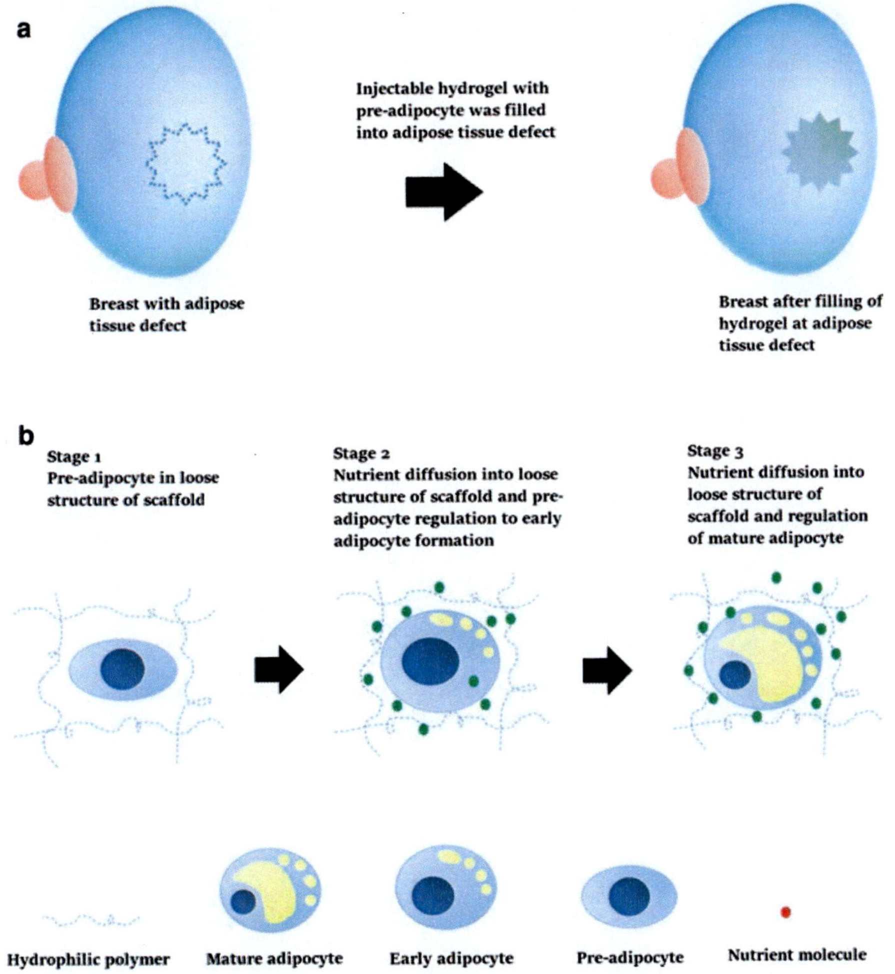


**Fig. 6.3** **a** Injection of hydrogel, incorporated with pre-adipocyte to the defect site and **b** response of adipocyte on injected hydrogel scaffold at the defect site

Generally, each type of tissue formation needs a different mechanical clue to regulate cells into different morphogenesis. For instance, bone tissue needs high modulus and strength. Furthermore, it also needs low elasticity and elongation [8–30]. Aorta needs low modulus and strength; it also prefers high elasticity and elongation [31] (Fig. 6.6).

Therefore, to mimic the mechanical performance, scaffolds can be engineered to be similar to different types of native tissue.

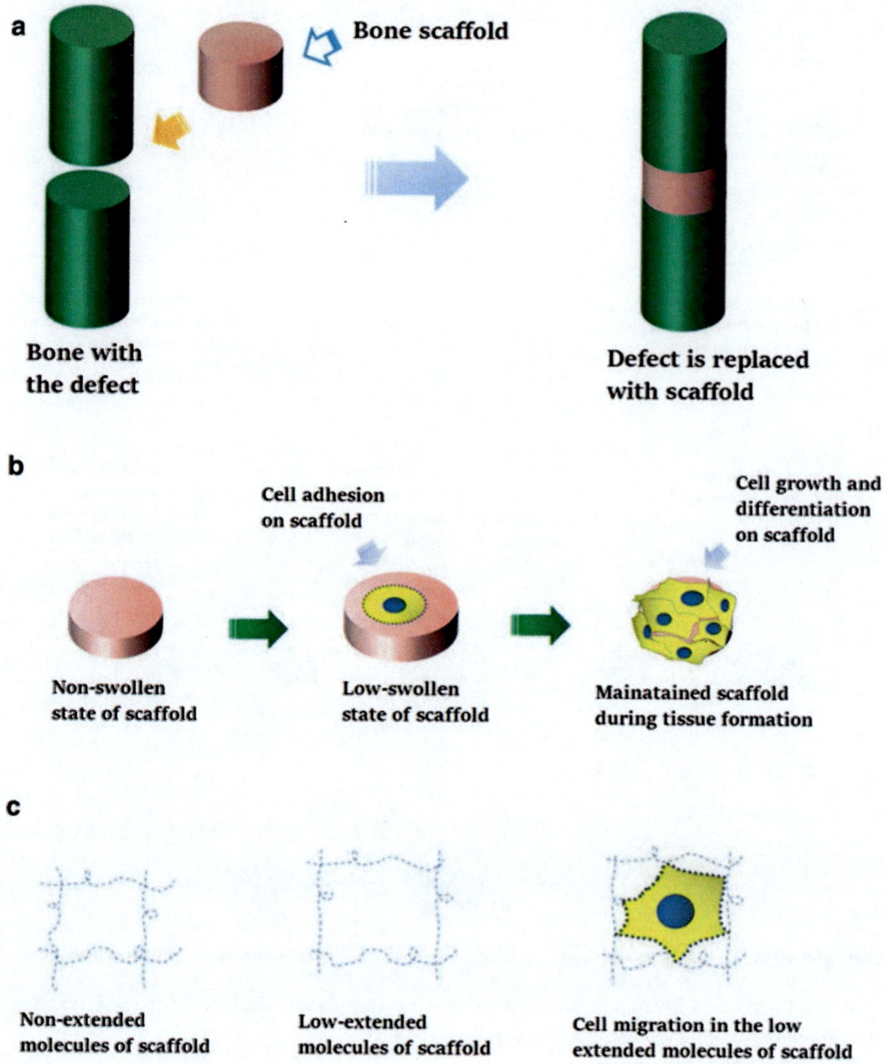


**Fig. 6.4** **a** Scaffold at the defect site of bone and **b** response of bone cells on scaffolds at the defect site

## 6.3 Approaches in Mimicking of Physical and Mechanical Function

There are many approaches to mimic mechanical functionalities, and these approaches are classified into (1) physical and chemical methods. Mainly, the approaches in mimicking of physical and mechanical functions are used to organize the molecules in the scaffold into different regularities. This has an effect on

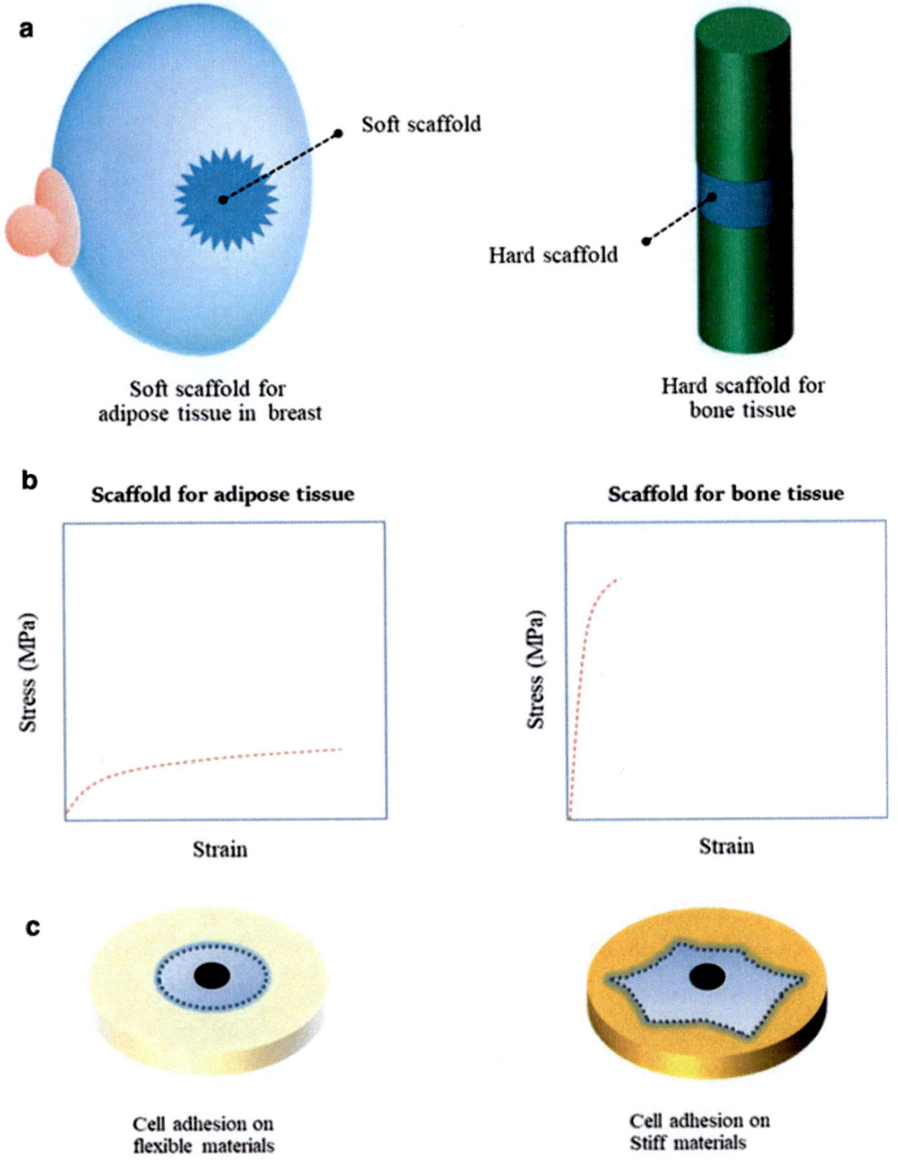


**Fig. 6.5** **a** Scaffolds at defect of breast and bone, **b** mechanical stress–strain graph of scaffold for adipose and bone tissue, and **c** cell adhesion on flexible and stiff materials

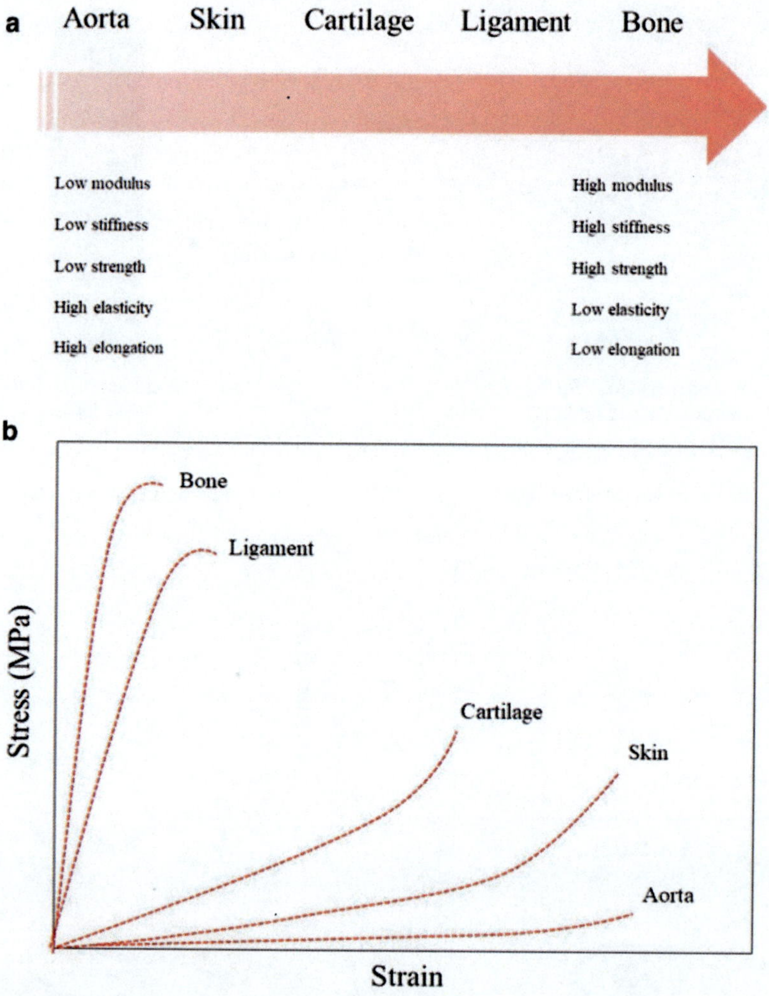


**Fig. 6.6** **a** Mechanical behaviors of tissues and **b** mechanical stress–strain graph of tissues

the different physical and mechanical functions of scaffolds that are similar to the different tissues.

### *6.3.1 Physical Approach for Molecular Organization in Mimicking*

The physical approach is used to organize polymer molecules of the scaffold into a regular structural formation. This approach is divided into two methods: (1) heat treatment and (2) blending, which are often used to organize the molecular structure in scaffolds.

#### 6.3.1.1 Heat Treatment for Molecular Organization

For heat treatment, scaffolds are mainly annealed at a temperature to induce molecular relaxation of the polymers [32, 33]. For the molecular relaxation during annealing, molecules before relaxation firstly demonstrate an amorphous structure. After heat treatment, molecules show more movement and self-organization into a regular structure (Fig. 6.7).

The molecular organization, before and after annealing, has different functions. For instance, in the case of mechanical function, the amorphous structure before annealing shows the lower mechanical strength than the regular structure after annealing [34]. For mimicking of mechanical function, similar to natural tissue, the regular structure scaffolds show high mechanical strength fitting to bone tissue [35–37]. On the other hand, the amorphous structure, which shows low mechanical strength, is suitable for soft tissue [38]. In some cases, scaffolds are incorporated with stem cells for mimicking tissue formation similar to natural tissue [1, 39]. The mechanical function of scaffold has an effect on stem cell regulation [1, 40, 41]. For instance, scaffolds with high modulus or stiffness show the clues to regulate stem cells into osteoblast [42, 43]. In contrast, scaffolds with low modulus or softness regulate stem cells into adipocyte or the other cells of soft tissue [42, 44].

For physical function, scaffolds with amorphous structures demonstrate a high swelling behavior, which is suitable for diffusion of nutrients. This is suitable to promote the formation for some tissues [45, 46]; for example, adipocytes require high amounts of nutrients to form adipose tissue. Therefore, the scaffolds which have high swelling are suitable for adipose tissue [45].

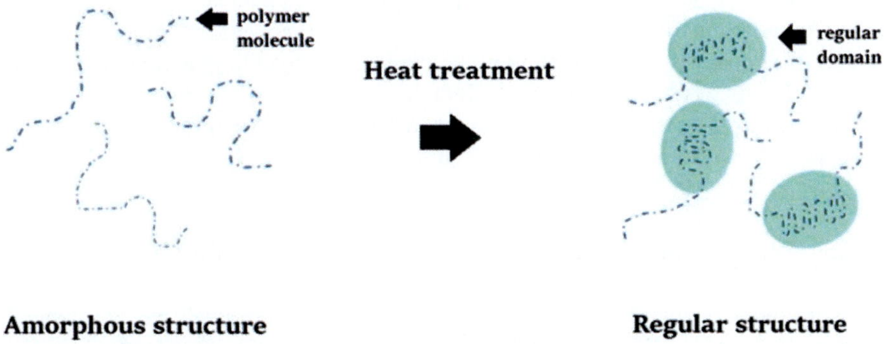


**Fig. 6.7** Molecular organization of polymers under heat treatment

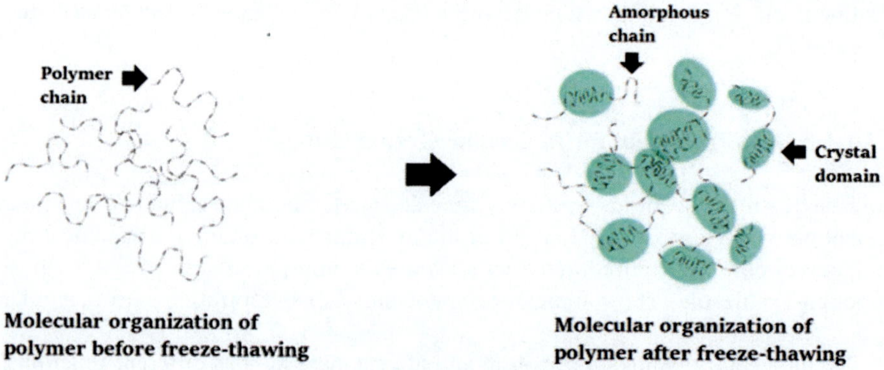


**Fig. 6.8** Molecular organization of polymers before and after freeze thawing

#### 6.3.1.2 Freeze Thawing for Molecular Organization

Freeze thawing is another physical approach that is used to organize molecules of polymers in scaffolds [47, 48]. In this approach, molecules of polymers are organized into crystalline molecules. For instance, this approach is used with poly (vinyl alcohol) (PVA). This approach starts with an aqueous solution of PVA, which has been frozen to form ice crystal domains. This frozen solution shows ice crystal domains as a discontinuous phase, which distributes into the polymer matrix as a continuous phase. Then, this frozen solution is thawed until the ice crystal domains are molten. Afterward, the solution, with molten ice crystal domains, is frozen for a second time. During this step, the molten ice crystal domain is frozen before being thawed for a second time. The freeze thawing is continued for a number of certain cycles to form hydrogel, which have two parts of molecular organization: (1) crystal domains and (2) amorphous chains (Fig. 6.8).

This freeze thawing is used to mimic physical and mechanical functions in scaffolds, similar to natural tissue. For instance, mimicked scaffolds similar to high, soft tissue with properties of high water uptake require hydrogel with a molecular organization of a few crystal domains and a lot of amorphous chains. On the other hand, scaffolds mimicking hard tissue with low water uptake are required to have a high amount of crystal domains and a low amount of the amorphous chain.

#### 6.3.1.3 Polymer Blending for Molecular Organization in Scaffolds

Polymer blending is a physical approach used to mimic physical and mechanical functions in scaffolds. When using this approach, two types of polymers are blended to form a material which combines their properties (Fig. 6.9a). An example of this is polymers with elasticity, which are blended with polymers with stiffness to create a novel polymer with toughness [49, 50]. For polymer blending, molecular compatibility and miscibility are of importance, as they have an influence on the performance of polymer blends [51, 52]. When certain polymer blends are incompatible, the molecular characteristic often demonstrates different physical and chemical functionalities [53, 54]. These differences have an effect on the formation of heterogeneous structures [55, 56] (Fig. 6.9b). In the case of incompatibility, there are surface tensions or non-interdiffusion of molecules between the phases of the polymer blends, leading to the formation of segregation at the interface areas [57, 58]. This results in a heterogeneous structure of polymer blends: These heterogeneous structures can be observed in microscale. In the case of compatibility, polymers must have the same molecular characteristics in molecular flexibility, hydrophilicity, and hydrophobicity. When the polymers have the same molecular characteristics, these polymers can interact with each other via chemical functionality. The molecular interactions have an effect on the formation of compatibility in polymer blends [59, 60]. In some cases, polymer blends gain improved compatibility by adding compatibilizers [61, 62]. A compatibilizer is a molecule that has the unique molecular characteristic of interacting with both phases in the polymer blend [62, 63]. Compatibilizers act as bridges to reduce the surface tension between the phases of a polymer blend. Immiscible polymer blends with compatibilizers can form morphology with fused texture at interface area by microscope observation [64]. For characterizations at the molecular level, there is a separation in the molecules of the polymers at the interfacial zone. In this case, polymers form a molecular interaction, without non-molecular entanglements. In the case of miscibility, the molecules of the polymer diffuse and merge together [65]. This leads to molecular interaction and entanglement during blending [66, 67]. The miscible polymer blending demonstrates the homogeneous formation at the molecular level. Importantly, for miscibility the molecular characteristics of polymers for blending must be quite similar.

Different tissues have different mechanical functions, and this in turn has an effect on cell regulation into different tissues. In order to construct scaffolds that have different mechanical function, for cell regulation into various tissues, polymers that have different molecular structures, physical characteristics, and behaviors are blended for fabrication into different mechanical behaviors within scaffolds. As an example, long linear chain polymers that have high flexibility have often been used as scaffolds for cell regulation into tissue which had elasticity [68]. Stiff-chain polymers are suitable for cell regulation into tissue which exhibit high strength [69]. Blending the polymers can adjust the mechanical behavior to fit different types of tissue. Therefore, it is important to select suitable molecular structures for polymer blends, so as to mimic physical and mechanical functions in scaffolds.

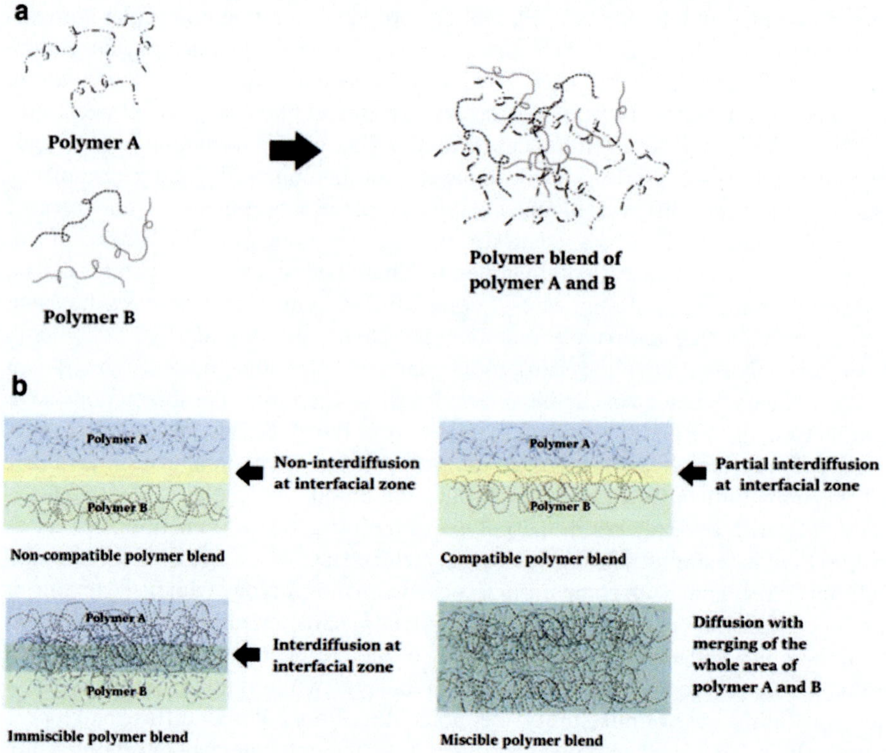


**Fig. 6.9** **a** Scenario of polymer blending and **b** molecular structural organization in polymer blending: non-compatible, compatible, immiscible, and miscible polymer blend

### 6.3.2 *Chemical Approaches for Molecular Organization in Mimicking*

For mimicking physical and mechanical function, a chemical approach is often used to create scaffolds which are similar to natural tissue. Chemical approaches have an advantage regarding the enhancement of both physical and mechanical stabilities of scaffolds. For the requirement of high stability during tissue formation, a chemical approach is suitable to mimic physical and mechanical functions in scaffolds. Chemical treatment for molecular crosslinking and transformation is the approach which is always used in scaffolds.

#### 6.3.2.1 Chemical Treatment for Molecular Crosslinking in Scaffolds

Chemical crosslinking fit to scaffolds based on polymers. Chemical, crosslinked scaffolds demonstrate the covalent bonding between their polymer molecules (Fig. 6.10).

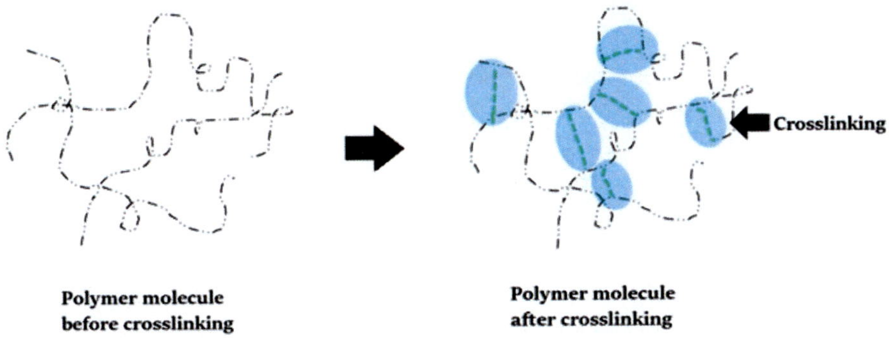


**Fig. 6.10** Chemical crosslinking of polymer molecules in scaffolds

For chemical crosslinking, there are two methods: (1) with and (2) without crosslinkers [70, 71]. For the first method, chemical, reactive molecules are used to crosslink with the polymer. These molecules have at least two chemical reactive groups in their structure. These groups form covalent bonding with the polymer molecules. For instance, glutaraldehyde (GA) is often selected as a crosslinker for protein. GA has two reactive groups of aldehyde on its molecule. For crosslinking based on GA, aldehyde groups form covalent bonding with amine groups (−NH3) and hydroxyl (−OH) of proteins. In the case of the second method, polymer molecules have to show chemical reactive groups that form covalent bonding with energy activation. For instance, polymer molecules are exposed to energy. This leads to the free radical generation in molecules of the polymers. Then, the free radical forms covalent bonding between molecules of polymers. Another example is a polymer with hydroxyl (−OH) and carboxylic (−COOH) groups in its molecules, which is activated with energy or chemical reactive condition. Then, the polymer forms crosslinking via covalent bonding of –OH and –COOH groups between its molecules.

Based on the chemical approach, these are used to mimic the physical and mechanical functions in scaffolds [72, 73]. For instance, scaffolds which need low water uptake and high mechanical strength are crosslinked with a high amount of crosslinkers. For low water uptake, these crosslinked scaffolds show the molecular structure with a dense network. This has an effect on decreasing water uptake in their molecular structure. In the case of high mechanical strength, the dense network demonstrates high mechanical strength. Low water uptake and high mechanical strength are functions which are mimicked for bone tissue. On the other hand, polymers which have a loose network of molecular structures, from low crosslinking, show physical and mechanical functions of high water uptake and low mechanical strength, respectively. These functions are mimicked to soft tissue, for instance skin and adipose tissue.

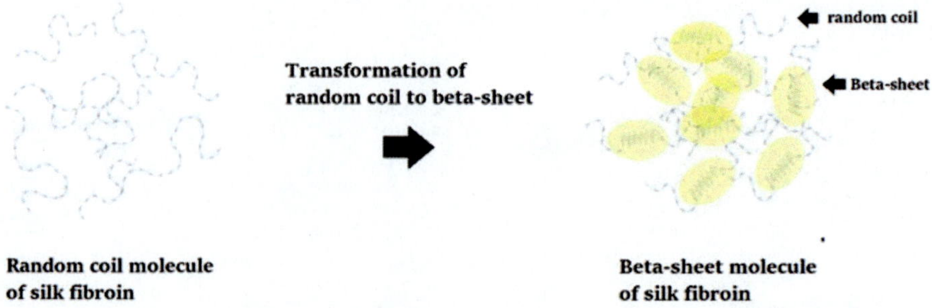


**Fig. 6.11** Transformation of random coil molecules to the beta sheet of silk fibroin

#### 6.3.2.2 Chemical Treatment for Molecular Transformation in Scaffolds

Chemical treatment for molecular transformation is an approach which is used to mimic scaffolds similar to physical and mechanical functions of natural tissue. For this approach, random coil molecules of polymers are transformed into beta sheet molecules of polymers [74]. For instance, a silk fibroin solution is casted into the mold to form the scaffold. After formation into scaffold, molecules of this silk fibroin scaffold have main random coil formation. Then, the scaffold is immersed into methanol or ethanol to transform these random coils to beta sheets of silk fibroin (Fig. 6.11).

This transformation of silk fibroin is used to mimic physical and mechanical function, similar to natural tissue [75]. For example, in the case of high mechanical strength tissues, silk fibroin has to transform into high amounts of beta sheet. This has an effect on the enhancement of its mechanical strength within the scaffolds, mimicking of that tissue.

Contained within this chapter is the explained, mimicked physical and mechanical functions in scaffolds.. The physical and chemical approaches which are used to mimic function in scaffold are described. Instances of mimicked physical and mechanical function, which are similar to natural tissue, are shown as guidance to create scaffolds and to promote tissue formation. Nevertheless, to complete mimicked scaffolds, which have the function similar to natural tissue, the biological function must be undertaken. This will be explained in the next chapter.

## References

1. Jaipaew, J., Wangkulangkul, P., Meesane, J., Raungrut, P., Puttawibul, P.: Mimicked cartilage scaffolds of silk fibroin/hyaluronic acid with stem cells for osteoarthritis surgery: Morphological, mechanical, and physical clues. Mater. Sci. Eng. C—Mater. **64**, 173–182 (2016)
2. Sang, L., Luo, D., Xu, S., Wang, X., Li, X.: Fabrication and evaluation of biomimetic scaffolds by using collagen–alginate fibrillar gels for potential tissue engineering applications. Mater. Sci. Eng. C—Mater **31**, 262–271 (2011)
3. Zhang, K., Song, L., Wang, J., Yan, S., Li, G., Cui, L., Yin, J.: Strategy for constructing vascularized adipose units in poly(l-glutamic acid) hydrogel porous scaffold through inducing

in-situ formation of ASCs spheroids. Acta Biomater. **51**, 246–257 (2017)
4. Kreimendahl, F., Köpf, M., Thiebes, A.L., Campos, D.F.D., Blaeser, A., Schmitz-Rode, T., Apel, C., Jockenhoevel, S., Fischer, H.: Three-dimensional printing and angiogenesis: tailored agarose-type I collagen blends comprise three-dimensional printability and angiogenesis potential for tissue-engineered substitutes. Tissue Eng. C-Me **23**, 604–615 (2017)
5. Agarwal, T., Kumar, T., Sudip, M., Ghosh, K.: Decellularized caprine liver-derived biomimetic and pro-angiogenic scaffolds for liver tissue engineering. Mater. Sci. Eng. C-Mater. **98**, 939–948 (2019)
6. Wangkulangkul, P., Jaipaew, J., Puttawibul, P., Meesane, J.: Constructed silk fibroin scaffolds to mimic adipose tissue as engineered implantation materials in post-subcutaneous tumor removal. Mater. Design **106**, 428–435 (2016)
7. Kreimendahl, F., Köpf, M., Thiebes, A.L., Campos, D.F.D., Blaeser, A., Schmitz-Rode, T., Apel, C., Jockenhoevel, S., Fischer, H.: Three-dimensional printing and angiogenesis: tailored agarose-type I collagen blends comprise three-dimensional printability and angiogenesis potential for tissue-engineered substitutes. Tissue Eng. C-Me **23**, 604–615 (2017)
8. Asadi, N., Alizadeh, E., Bakhshayesh, A.R.D., Mostafavi, E., Akbarzadeh, A., Soodabeh, D.S.: Fabrication and in vitro evaluation of nanocomposite hydrogel scaffolds based on gelatin/PCL–PEG–PCL for cartilage tissue engineering. ACS Omega **4**(1), 449–457 (2019)
9. Tsai, M.C., Hung, K.C., Hung, S.C., Hsu, S.H.: Evaluation of biodegradable elastic scaffolds made of anionic polyurethane for cartilage tissue engineering. Colloids Surf. B Biointerf. **125**, 34–44 (2015)
10. Farokhi, M., Shariatzadeh, F.J., Solouk, A., Hamid, M.H.: Alginate based scaffolds for cartilage tissue engineering: a review. Int. J. Polym. Mater. Po **69**, 230–247 (2020)
11. Flynn, L., Semple, J.L., Woodhouse, K.A.: Decellularized placental matrices for adipose tissue engineering. J. Biomed. Mater. Res. A **79A**, 359–369 (2006)
12. Lee, S., Lee, H.S., Chung, J.J., Kim, S.H., Park, J.W., Lee, K., Jung, Y.: Enhanced regeneration of vascularized adipose tissue with dual 3D-printed elastic polymer/dECM hydrogel complex. Int. J. Mol. Sci. **22**, 2886 (2021)
13. Negrini, N.C., Tarsini, P., Tanzi, M.C., Farè, S.: Chemically crosslinked gelatin hydrogels as scaffolding materials for adipose tissue engineering. J. Appl. Polym. Sci. **136**, 47104 (2019)
14. Chang, K.H., Liao, H.T., Chen, J.P.: Preparation and characterization of gelatin/hyaluronic acid cryogels for adipose tissue engineering: In vitro and in vivo studies. Acta Biomater. **9**, 9012–9026 (2013)
15. Matinfar, M., Mesgar, A.S., Mohammadi, Z.: Evaluation of physicochemical, mechanical and biological properties of chitosan/carboxymethyl cellulose reinforced with multiphasic calcium phosphate whisker-like fibers for bone tissue engineering. Mater. Sci. Eng. C-Mater. **100**, 341–353 (2019)
16. Chen, Y., Kawazo, N., Chen, G.: Preparation of dexamethasone-loaded biphasic calcium phosphate nanoparticles/collagen porous composite scaffolds for bone tissue engineering. Acta Biomater. **67**, 341–353 (2018)
17. Nie, L., Chen, D., JSuo, J., Zou, P., Feng, S., Yang, Q., Yang, S., Ye, S.: Physicochemical characterization and biocompatibility in vitro of biphasic calcium phosphate/polyvinyl alcohol scaffolds prepared by freeze-drying method for bone tissue engineering applications. Colloid Surf. B **100**, 169–176 (2012)
18. Narayan, B., Vivek, S., Sarada, V., Mallick, P., Jain, Y., Sinh, S., Rastogi, A., Srivastava, P.: Design and evaluation of chitosan/chondroitin sulfate/nano-bioglass based composite scaffold for bone tissue engineering. Int. J. Biol. Macromol. **133**, 817–830 (2019)
19. Singh, B.N., Veeresh, V., Mallick, S.P., Sinh, S., Rastogi, A., Srivastava, P.: Generation of scaffold incorporated with nanobioglass encapsulated in chitosan/chondroitin sulfate complex for bone tissue engineering. Inter J Biol Macromol **153**, 1–16 (2020)
20. Olad, A., Hagh, H.B.K., Mirmohseni, A., Azhar, F.F.: Graphene oxide and montmorillonite enriched natural polymeric scaffold for bone tissue engineering. Ceram Inter. **45**, 15609–15619 (2019)

21. Jaipaew, J, Wangkulangkul, P., Meesane, J., Raungrut, P., Puttawibul, P.: Mimicked cartilage scaffolds of silk fibroin/hyaluronic acid with stem cells for osteoarthritis surgery: morphological, mechanical, and physical clues. Mater. Sci. Eng. C-Mater. **64**, 173–182 (2016)
22. Banerjee, A., Arha, M., Choudhary, S., Ashton, R.S., Bhatia, S.R., Schaffer, D.V., Kane, R.S.: The influence of hydrogel modulus on the proliferation and differentiation of encapsulated neural stem cells. Biomaterials **30**, 4695–4699 (2009)
23. Griffin, M.F., Butler, P.E., Seifalian, A.M., Kalaskar, D.M.: Control of stem cell fate by engineering their micro and nanoenvironment. World J. Stem Cells **7**(1), 37–50 (2015)
24. Davidenkoa, N., Campbell, J.J., Thiana, E.S., Watson, C.J., Cameron, R.E.: Collagen–hyaluronic acid scaffolds for adipose tissue engineering. Acta Biomater. **6**, 3957–3968 (2010)
25. Frydrych, M., Román, S., MacNeil, S., Chen, B.: Biomimetic poly(glycerol sebacate)/poly(l-lactic acid) blend scaffolds for adipose tissue engineering. Acta Biomater. **18**, 40–49 (2015)
26. Gleeson, J.P., Plunkett, N.A., O'Brien, F.J.: Addition of hydroxyapatite improves stiffness, interconnectivity and osteogenetic potential of a highly porous collagen-based scaffold for bone tissue regeneration. Eur. Cells. Mater. **20**, 218–230 (2010)
27. Xu, C., Su, P., Chen, X., Meng, C., Yu, W., Xiang, A.P., Wang, Y.: Biocompatibility and osteogenesis of biomimetic Bioglass-Collagen-Phosphatidylserine composite scaffolds for bone tissue engineering. Biomaterials **32**, 1051–1058 (2011)
28. Hung, B.P., Hutton, D.L., Grayson, W.L.: Mechanical control of tissue-engineered bone. Stem Cell. Res. Ther. **4**, 10 (2013)
29. Breuls, R.G.M., Jiya, T.U., Smit, T.H.: Scaffold stiffness influences cell behavior: opportunities for skeletal tissue engineering. Open Orthop. J. **2**, 103–109 (2008)
30. Engler, A.J., Sen, S., Sweeney, H.L., Discher, D.E.: Matrix elasticity directs stem cell lineage specification. Cell **126**(4), 677–689 (2006)
31. Coenen, A.M.J., Bernaerts, K.V., Harings, J.A.W., Jockenhoevel, S., Ghazanfari, S.: Elastic materials for tissue engineering applications: Natural, synthetic, and hybrid polymers. Acta Biomater. **79**, 60–82 (2018)
32. Makrani, N., Ammari, A., Benrekaa, N., Rodrigu, D., Giroux, Y.: Dynamics of the α-relaxation during the crystallization of PLLA and the effect of thermal annealing under humid atmosphere. Polym. Degrad. Stabil. **164**, 90–101 (2019)
33. Nguyen, H.K., Kawaguchi, D., Tanaka, K.: Effect of molecular architecture on conformational relaxation of polymer chains at interfaces. Macromol. Rapid Comm. **41**, 2000096 (2020)
34. Abhari, R.E., Mouthuy, P.A., Zargar, N., Brown, C., Carr, A.: Effect of annealing on the mechanical properties and the degradation of electrospun polydioxanone filaments. J. Mech. Behav. Biomed. **67**, 127–134 (2017)
35. Ozdil, D., Aydin, H.M.: Polymers for medical and tissue engineering applications. J. Chem Technol. Biot. **89**, 1793–1810 (2014)
36. Onyishi, H.O., Oluah, C.K.: Effect of stretch ratio on the induced crystallinity and mechanical properties of biaxially stretched PET. Phase Tran. **93**, 924–934 (2020)
37. Collins, M.N., Ren, G., Young, K., Pina, S., Reis, R.L., Oliveira, J.M.: Scaffold fabrication technologies and structure/function properties in bone tissue engineering. Adv. Funct. Mater. **31**, 2010609 (2021)
38. Sang, Y., Li, M., Liu, J., Yao, Y., Ding, Z., Wang, L., Xiao, L., Lu, Q., Fu, X., Kaplan, D.L.: Biomimetic silk scaffolds with an amorphous structure for soft tissue engineering. ACS Appl. Mater. Interf. **10**(11), 9290–9300 (2018)
39. Cheung, H.K., Han, T.T.Y., Marecak, D.M., Watkins, J.F., Amsden, B.G., Flynn, L.E.: Composite hydrogel scaffolds incorporating decellularized adipose tissue for soft tissue engineering with adipose-derived stem cells. engineering with adipose-derived stem cells. Biomaterials **35**, 1914–1923 (2014)
40. Murphy, C.M., Matsiko, A., Haugh, M.G., Gleeson, J.P., O'Brien, F.J.: Mesenchymal stem cell fate is regulated by the composition and mechanical properties of collagen–glycosaminoglycan scaffolds. J. Mech. Behav. Biomed. **11**, 53–62 (2012)

41. Yao, R., He, J., Meng, G., Jiang, B., Wu, F.: Electrospun PCL/Gelatin composite fibrous scaffolds: mechanical properties and cellular responses. J. Biomat. Sci-Polym. E **27**, 824–838 (2016)
42. Chen, G., Dong, C., Yang, L., Lv, Y.: 3D scaffolds with different stiffness but the same microstructure for bone tissue engineering. ACS Appl. Mater. Interfaces **7**(29), 15790–15802 (2015)
43. Murphy, C.M., Matsiko, A., Haugh, M.G., Gleeson, J.P., O'Brien, F.J.: Mesenchymal stem cell fate is regulated by the composition and mechanical properties of collagen–glycosaminoglycan scaffolds. J. Mech. Behav. Biomed. Mater. **11**, 53–62 (2012)
44. Young, D.A., Choi, Y.S., Engler, A.J., Christman, K.L.: Stimulation of adipogenesis of adult adipose-derived stem cells using substrates that mimic the stiffness of adipose tissue. Biomaterials **34**, 8581–8588 (2013)
45. Von Heimburg, D., Kuberk, M., Rendchen, R., Hemmrich, K., Rau, G., Pallua, N.: Preadipocyte-loaded collagen scaffolds with enlarged pore size for improved soft tissue engineering. Int. J. Artif. Organs **26**, 1064–1076 (2003)
46. Zhang, F., He, C., Cao, L., Feng, W., Wang, H., Mo, X., Wang, J.: Fabrication of gelatin–hyaluronic acid hybrid scaffolds with tunable porous structures for soft tissue engineering. Int. J. Biol. Macromol. **48**, 474–481 (2011)
47. Gupta, S., Webster, T.J., Sinha, A.: Evolution of PVA gels prepared without crosslinking agents as a cell adhesive surface. J. Mater. Sci. Mater. Med. **22**, 1763–1772 (2011)
48. Gupta, S., Goswami, S., Sinha, A.: A combined effect of freeze—thaw cycles and polymer concentration on the structure and mechanical properties of transparent PVA gels. Biomed Mater. **7**(1), 015006 (2012)
49. Wan, H., Shen, J., Gao, N., Liu, J., Gao, Y., Zhang, L.: Tailoring the mechanical properties by molecular integration of flexible and stiff polymer networks. Soft Matt. **14**, 2379–2390 (2018)
50. Peng, M., Xiao, G., Tang, X., Zhou, Y.: Hydrogen-Bonding Assembly of Rigid-Rod Poly(p-sulfophenylene terephthalamide) and Flexible-Chain Poly(vinyl alcohol) for Transparent, Strong, and Tough Molecular Composites. Macromolecules **47**(23), 8411–8419 (2014)
51. Askadskii, A.A., Matseevich, T.A., Popova, M.N., Kondrashchenko, V.I.: Prediction of the compatibility of polymers and analysis of the microphase compositions and some properties of blends. Polym. Sci. Ser. A **57**, 186–199 (2015)
52. Zhang, J., Wang, Z., Wang, Q., Ma, J., Cao, J., Hu, W., Wu, Z.: Relationship between polymers compatibility and casting solution stability in fabricating PVDF/PVA membranes. J. Membrane Sci. **537**, 263–271 (2017)
53. Beattie, D.L., Mykhaylyk, O.O., Armes, S.P.: Enthalpic incompatibility between two steric stabilizer blocks provides control over the vesicle size distribution during polymerization-induced self-assembly in aqueous media. Chem. Sci. **11**, 10821–10834 (2020)
54. Zaikin, A.E., Bobrov, G.B.: Compatibilization of blends of incompatible polymers via filling. Polym. Sci. Ser. A **54**, 651–657 (2012)
55. Dobrovszky, K., Ronkay, F.: Effects of phase inversion on molding shrinkage, mechanical, and burning properties of injection-molded PET/HDPE and PS/HDPE polymer blends. Polym Plast Technol. Eng. **56**, 1147–1157 (2017)
56. Torquato, S.: Disordered hyperuniform heterogeneous materials. J. Phys. Condens Matt. **28**, 414012 (2016)
57. Wang, J., Tsou, A.H., Favis, B.D.: Effects of polyethylene molecular weight distribution on phase morphology development in poly(p-phenylene ether) and polyethylene blends. Macromolecules **51**(22), 9165–9176 (2018)
58. Hu, K., Huang, D., Jiang, H., Sun, S., Ma, Z., Zhang, K., Pan, L., Li, Y.: Toughening biosourced poly(lactic acid) and poly(3-hydroxybutyrate-co-4-hydroxybutyrate) blends by a renewable poly(epichlorohydrin-co-ethylene oxide) elastomer. ACS Omega **4**(22), 19777–19786 (2019)
59. Lin, Q., Zheng, X., Gu, X., Zhao, L., Li, J., Li, Y.: Reactive splicing compatibilization of immiscible polymer blends: Compatibilizer synthesis in the melt state and compatibilizer architecture effects. Polymer **185**, 121952 (2019)

60. Shamsuri, A.A., Jamil, S.N.A.M.: Compatibilization effect of ionic liquid-based surfactants on physicochemical properties of PBS/rice starch blends: an initial study. Mat. **13**(8), 1885 (2020)
61. Ahmadlouydara, M., Chamkouri, M., Chamkouri, H.: Compatibilization of immiscible polymer blends (R-PET/PP) by adding PP-g-MA as compatibilizer: analysis of phase morphology and mechanical properties. Polym. Bull. **77**, 5753–5766 (2020)
62. Yang, X., Wang, H., Chen, J., Fu, Z., Zhao, X., Li, Y.: Copolymers containing two types of reactive groups: New compatibilizer for immiscible PLLA/PA11 polymer blends. Polymer **177**, 139–148 (2019)
63. Ding, Y., Feng, W., Huang, D., Lu, B., Wang, P., Wang, G., Ji, J.: Compatibilization of immiscible PLA-based biodegradable polymer blends using amphiphilic di-block copolymers. Eur. Polym. J. **118**, 45–52 (2019)
64. Seier, M., Stanic, S., Koch, T., Archodoulaki, V.M.: Effect of different compatibilization systems on the rheological, mechanical and morphological properties of polypropylene/polystyrene blends. Polymers **12**, 2335 (2020)
65. Colmenero, J.: Polymer chain diffusion in polymer blends: A theoretical interpretation based on a memory function formalism. J. Polym. Sci. Part B Polym. Phys. **57**, 1239–1245 (2019)
66. Fekete, E., Földes, E., Pukánszky, B.: Effect of molecular interactions on the miscibility and structure of polymer blends. Eur. Polym. J. **41**, 727–736 (2005)
67. Wu, B., Cai, Y., Zhao, X., Ye, L.: Fabrication of well-miscible and highly enhanced polyethylene/ultrahigh molecular weight polyethylene blends by facile construction of interfacial intermolecular entanglement. Polym. Test. **93**, 106973 (2021)
68. Parrag, I.C., Woodhouse, K.A.: Development of biodegradable polyurethane scaffolds using amino acid and dipeptide-based chain extenders for soft tissue engineering. J. Biomat. Sci-Polym. E **21**, 843–862 (2010)
69. Ying, T.H., Ishii, D., Mahara, A., Murakami, S., Yamaoka, T., Sudesh, K., Samian, R., Fujita, M., Maeda, M., Iwata, T.: Scaffolds from electrospun polyhydroxyalkanoate copolymers: Fabrication, characterization, bioabsorption and tissue response. Biomaterials **29**, 1307–1317 (2008)
70. Duquette, D., Dumont, M.J.: Comparative studies of chemical crosslinking reactions and applications of bio-based hydrogels. Polym. Bull. **76**, 2683–2710 (2019)
71. Metters, A., Hubbell, J.: Network formation and degradation behavior of hydrogels formed by Michael-type addition reactions. Biomacromol. **6**(1), 290–301 (2005)
72. Improvements in mechanical properties of collagen-based scaffolds for tissue engineering. Curr. Opin. Biomed. Eng. **17**, 100253 (2021)
73. Haugh, M.G., Murphy, C.M., McKiernan, R.C., Altenbuchner, C., O'Brien, F.J.: Crosslinking and mechanical properties significantly influence cell attachment, proliferation, and migration within collagen glycosaminoglycan scaffolds. Tissue Eng. A **77**, 1201–1208 (2011)
74. Yang, S.Y., Kim, S., Shin, H., Choi, S.H., Kim, Y.L., Joo, C., Ryu, W.H.: Random lasing detection of structural transformation and compositions in silk fibroin scaffolds. Nano Res. **12**, 289–297 (2019)
75. Koh, L.D., Cheng, Y., Teng, C.P., Khin, Y.W., Loh, X.J., Tee, S.Y., Low, M., Ye, E., Yu, H.D., Zhang, Y.W., Han, M.Y.: Structures, mechanical properties and applications of silk fibroin materials. Prog. Polym. Sci. **46**, 86–110 (2015)

# Chapter 7
# Mimicked Biological Function of Scaffolds

The biological function of scaffolds plays an important role in the promotion of tissue formation [1, 2]. Especially, biodegradation and bioactivity are important functions for cell regulation, for completion of the tissue formation [3, 4]. Mimicking of biodegradation and bioactivity during tissue formation is the approach to guide the cells to regulate themselves into the completed tissue formation [1, 5, 6]. This chapter demonstrates the mimicking of biological function: biodegradability and bioactivity, during tissue formation to create scaffolds. Some instances, as well as ideas, are shown to be guidance to design scaffolds with abiological function similar to that of natural tissue.

## 7.1 Mimicking of the Biological Function in Scaffolds

Mimicking biological functions mainly focuses on a microenvironment: the extracellular matrix (ECM) and growth factors, which show degradability and bioactivity to promote tissue formation [7, 8]. For ECMs, they demonstrate degradability during tissue formation [9, 10], as they are digested by enzymes of the matrix metalloproteinase (MMP). This leads to the loose structure of digested ECMs, which has an effect on the enhancement of cell behaviors: proliferation, migration, and differentiation. These cell behaviors promote tissue formation. Furthermore, molecules of ECMs have amino acid motif, which shows functionality on cells [11–13]. This motif acts as binding sites and biological signals for cell recognition and response, and in turn promotes cell behaviors: adhesion, proliferation, spreading, migration, and differentiation; all of which leads to enhancement of the tissue formation.

For growth factors, they mainly show functionality, which promotes the cell behaviors to induce tissue formation [14, 15]. Growth factors have amino acid motif, which acts as binding site and biological signals to activate cell behaviors: adhesion, proliferation, migration, and differentiation for inducing of tissue formation. Based on

 
J. Meesane, *Mimicked Tissue Engineering Scaffolds for Maxillofacial and Articular Cartilage Surgery*, Engineering Materials,
https://doi.org/10.1007/978-981-19-7830-2_7

the potential to promote tissue formation, the biodegradability and bioactivity of the microenvironment is used as guidance to create mimicked biological functional scaffolds.

### 7.1.1 Mimicking of Degradation During Tissue Formation

To mimic the function of degradation, similar to natural tissue, scaffolds are created with two main types of degradation during tissue regeneration: (1) hydrolytic degradation and (2) degradation by enzymes [16, 17]. Hydrolytic degradation begins with water absorption of the scaffolds that leads to chain scission. Chain scission decreases the molecular weight that leads to the degradation of scaffolds (Fig. 7.1a). Degradation by enzymes occurs when the enzymes attach to the scaffolds and digest the texture of the scaffolds. The digestion of scaffolds leads to degradation (Fig. 7.1b). In the design of performance scaffolds, the rate of degradation has to fit the rate of new tissue regeneration [18].

To mimic degradable functions, in the case of hydrolytic degradation, hydrophilic polymers, natural and synthetic polymers, are usually used to create scaffolds for tissue engineering [18–20]. Hydrophilic polymers can enhance water absorption that has an effect on degradation [21]. Degradation depends on the chemical functional groups that are caged around by water molecules [21], so the number of functional groups has an influence on the rate of degradation. In the case of a high number of functional groups, degradation of the scaffolds occurs at a high rate, whereas a low number of functional groups reduces the rate of degradation [22, 23].

For mimicking of degradation by enzyme, molecules, with degradable domains, are often selected as base materials to create scaffolds [23, 24]. These molecules are digested by enzymes at their degradable domains during tissue regeneration [25, 26]. The enzyme that often has an effect on digestion is MMP. Some scaffolds are created based on polymers incorporated with these biological molecules to mimic an enzyme degradable function [27].

This mimicking of biological function: hydrolytic degradation and degradation with enzymes, has an important role in the regulation of the tissue formation. Nevertheless, to complete mimicking of biological function, bioactivity of the microenvironment is undertaken.

### 7.1.2 Mimicking of Bioactivity for Enhancement of Tissue Formation

Bioactivity is the important function, in which it has the important role of inducing tissue formation [28, 29]. Especially, bioactivity of the microenvironment: ECMs and growth factors act as the biological signals to induce cell behaviors, leading to their

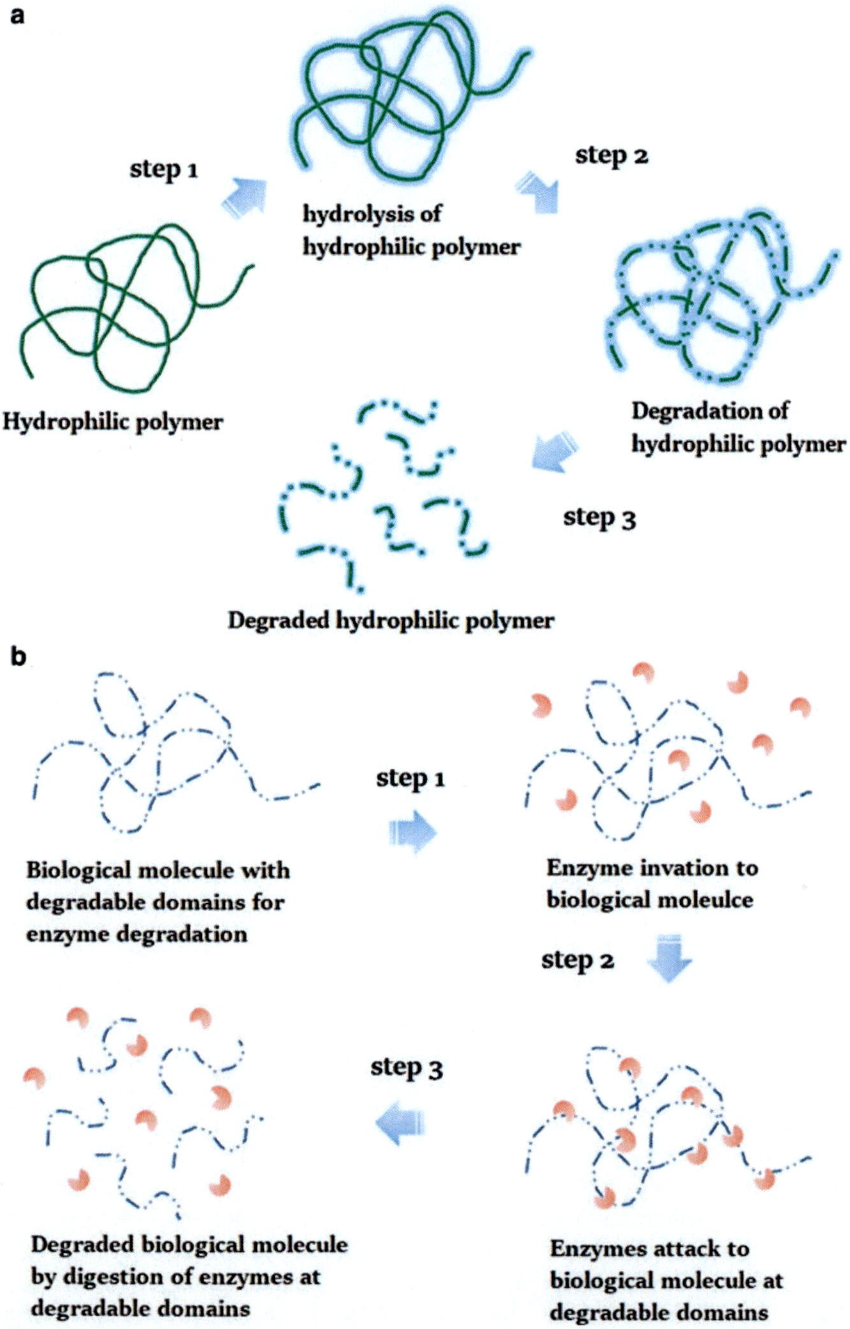


**Fig. 7.1** **a** Mechanism of hydrolytic degradation **b** Mechanism of degradation by metalloproteinase enzyme

promotion of tissue formations [30, 31]. Mimicking of the bioactivity of microenvironment is an approach for the creation of performance scaffolds to promote tissue formation. To mimic performance scaffolds, which have bioactivity similar to a microenvironment, polymers are firstly incorporated with components of ECMs, for instance collagens, glycosaminoglycans, and proteoglycans [32–34]. These components have bioactive function to promote cell behaviors to enhance tissue formation [35]. Secondly, polymers are combined with growth factors: bone morphogenetic proteins (BMPs), transforming growth factor-β (TGF-β), vascular endothelial growth factor (VEGF), and so forth. [36]. Growth factors are biological signals that activate cell behaviors to induce tissue formation [37]. Thirdly, polymers incorporated with ECMs and growth factors are used to create scaffolds that have biological functions close to a microenvironment [38].

## 7.2 Physical Approach in Mimicked Biological Function on Scaffolds

According to mimicking of biological function, degradability and bioactivity, scaffolds are fabricated by a physical approach, with irregular and regular formations of molecules. For irregular formations of molecules, polymers are generally mixed with biological signals: ECMs and growth factors, without adjusted condition [39]. Irregular formations are generated via molecular interaction without ordering. On the other hand, regular formations occur in adjusted conditions, which leads to self-organize or assembly of the molecules. The regular formation of molecules shows the ordering of interaction in the structure [40].

### *7.2.1 Physical Approach with an Irregular Formation of Molecules*

A physical approach with an irregular formation of molecules is a simple method to create the mimicked biological function of scaffolds. Using this approach, polymers are mixed with ECMs or growth factors, before being prepared into scaffolds [39]. The mixing is under non-adjusted conditions, and the molecules, in this condition, show irregular formation. For instance, in the case of mimicked biological function of bioactivity on cell behaviors, positive or negative charged polymers are mixed with biological molecules [41, 42]. Polymer molecules form physical interaction with the biological molecules via electrostatic forces, hydrophilic or hydrophobic interaction, and hydrogen bonding [43, 44]. This mixing shows a non-irregular formation, with an amorphous structure. The physical interaction from this mixing is weak, which can be broken in the environment during tissue formation [45, 46]. This leads to the releasing of biological molecules from polymer molecules to attach with cells

[46]. These scaffolds are suitable for rapid activation on cell behaviors at the early state of tissue formation [47]; however, they are non-suitable for any long period of activation on cell behavior for the latter state of tissue formation.

### *7.2.2 Physical Approach with Regular Formation of Molecules*

Using a physical approach, with a regular formation of molecules, polymers are mixed with ECMs or growth factors in the adjusted condition, which leads to self-assembly [48, 49]. This regular formation shows ordered structure which has stability in the environment of tissue formation [48, 49]. This structure has the important role of supporting bioactivity in the cells. The structural stability mainly has an effect on the enhancement on cell adhesion and proliferation, which leads to promoting tissue formation [50], and this can also prolong the bioactivity in cells [51]. For instance, the system of polymers mixed with collagen is adjusted in condition to form self-assembly [49, 52]. For this system, collagen is selected as the biological molecule to mimic biological function in the scaffolds. Polymers are used as base materials to enhance the physical stability and mechanical properties of scaffolds. In the adjusted condition, collagen self-assembly is similar to ECM. Self-assembled collagen shows a fibril structure that has stability for cell adhesion and proliferation. Nevertheless, self-assembled structures have physical interaction, which is broken in condition during tissue formation [53]. Therefore, to mimic biological function in scaffolds, a chemical approach is required to induce structural stability for some cases of tissue formation.

## 7.3 Chemical Approach in Mimicked Biological Function on Scaffolds

In the case of chemical approach, the biological molecules often connect to polymer molecules via covalent bonding. These polymers, connected with biological molecules, are difficult to break, as these biological molecules are fixed on the polymer molecule with stability. They perform as signals to activate cell behaviors, particularly in the later stages of tissue formation [54, 55]. Normally, there are two chemical approaches: (1) surface immobilization and (2) molecular grafting or conjugation, which are used to modify the mimicked biological function of scaffolds [56, 57].

### 7.3.1 *Surface Immobilization in Mimicked Biological Function on Scaffolds*

Surface immobilization, this approach is used to modify scaffold [58]. This scaffold must have reactive chemical groups to immobilize biological molecules on its surface [59]. In some cases, the scaffold has no reactive chemical groups in their molecules. This type of scaffold needs modification with reactive chemical groups, via chemical bonding, before immobilization of biological molecules [60]. For instance, the scaffold is modified with immobilization of growth factors on its surface [60] (Fig. 7.2a). Firstly, the surface of the scaffold is treated with chemical active molecules. Then, growth factors approach and attach themselves to the surface of the scaffolds, and these growth factors are immobilized via covalent bonding on the surface of scaffold.

Therefore, bioactive amino motifs of growth factors are immobilized on the surface of the scaffold. These motifs, for example, RGD and YIGSIR, act as the recognized site for cell attachment [61–63]. These motifs are usually immobilized on the surface of the scaffold via chemical bonding by spacers (Fig. 7.2b). For this immobilization, the spacers have two reactive groups that can generate chemical bonding with scaffolds. The spacers have effects on the biological function of cell behavior to the scaffolds. Long-chain spacers have more free movement to attach with cells than short-chain spacers [64]. This leads to non-stable attachment of cells on the surface of scaffolds that has an effect on decreasing tissue formation [64].

### 7.3.2 *Molecular Grafting or Conjugation in Mimicked Biological Function of Scaffolds*

Molecular grafting or conjugation is an approach of modification, which is used to mimic biological function of the microenvironment on scaffolds [65]. For molecular grafting, biological molecules are grafted with polymer molecules via chemical bonding [66]. Like as in immobilization, components of the ECM or growth factors are selected as base biological molecules for grafting with the polymer. These biological molecules are grafted with the chemical reactive side groups of polymer molecules [67–69] (Fig. 7.3). The molecular structural organization of grafted polymers has an effect on cell behavior, particularly in attachment via recognition on the binding site of biological molecules having an effect to promote cell attachment [70]. On the other hand, in the case of non-suitable amounts: hyper- or hypo-amounts of biological molecules act as inhibitors for cell activity [71, 72]. Another instance is grafted polymer molecules which self-organize into aggregation having lower cell attachment than non-aggregation. This is because the aggregation acts as a hindrance for cell activity [71, 72].

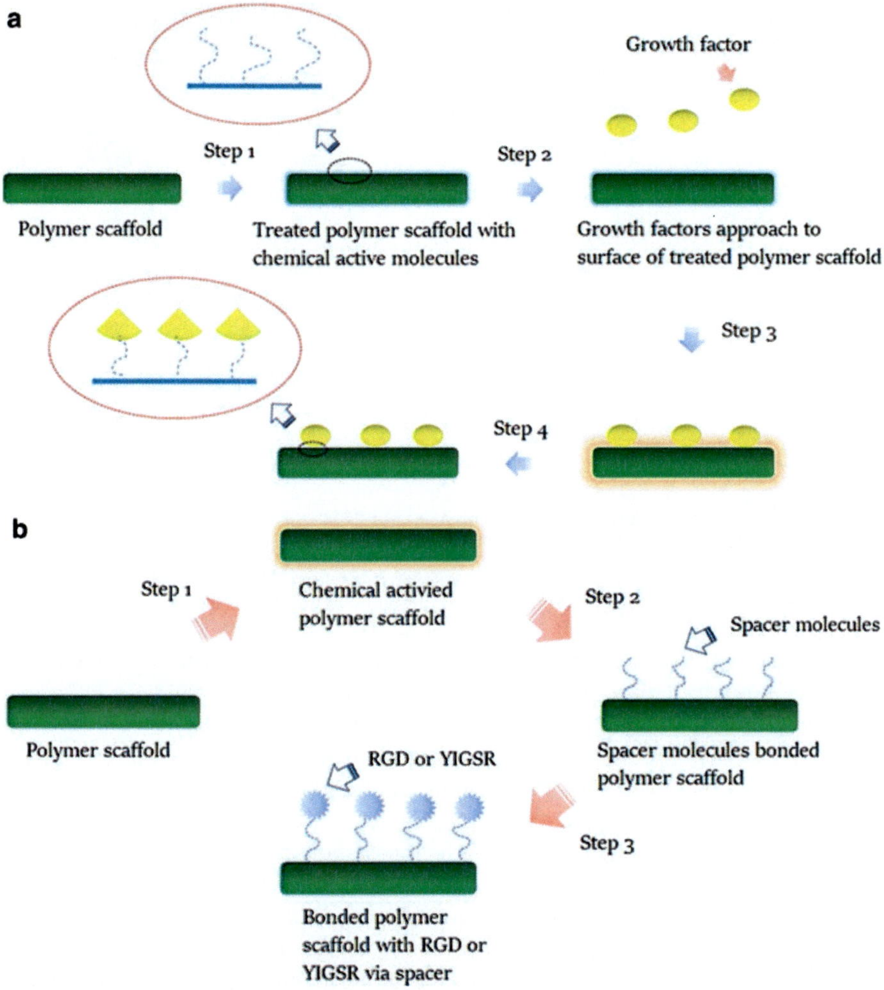


**Fig. 7.2** Modification of scaffolds by growth factors via chemical bonding. Grafted amino sequence on scaffolds via spacers

## 7.4 Design of Mimicked Biological Function in Scaffolds

Based upon (1) biological molecules, (2) selected polymers, and (3) a suitable approach, the mimicked biological function of scaffolds can be created with various designs. First, biological molecules are used with single or multi-components in the design of mimicking [73]. Second, degradable polymers: natural or synthetic polymers, are often selected as base materials to mimic scaffolds [74]. Third, either a physical or chemical approach is chosen as the approach for modification in scaffolds [75]. For instance, of a design of mimicked biological function in scaffolds,

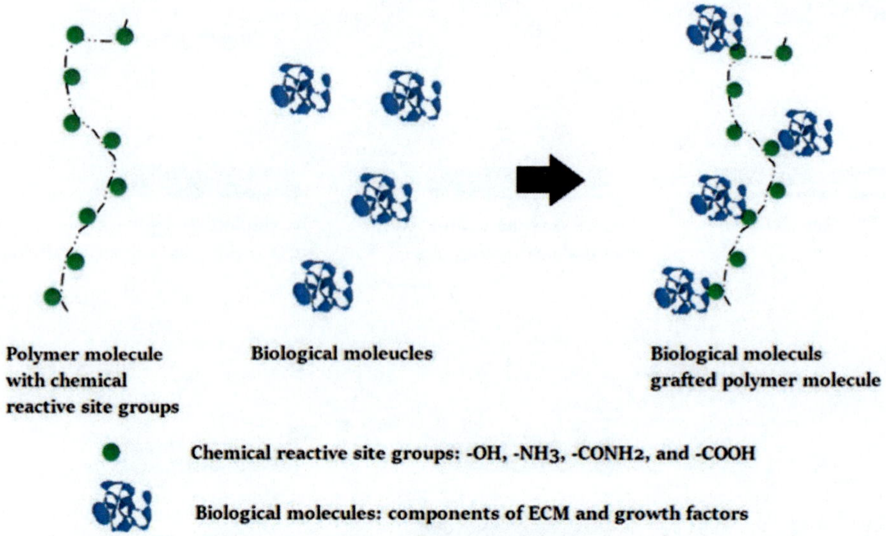


**Fig. 7.3** Molecular grafting of polymer molecules with biological molecules

bone regeneration needs BMPs as signals to induce bone formation and vascular endothelial growth factors (VEGFs) act as signals to induce new blood vessel formation [76]. Therefore, some researchers use BMPs and VEGFs to incorporate with the polymers, which can induce new bone tissue formation. Connecting these signals with polymers via this approach can mimic the programming of these signals for bone formation [76].

For more details of this base design, it is important to understand the behavior of signals in the design of programs of tissue regeneration similar to the native environment. The native environment has many different sequences of biological release from the cells in the regulation of the proposed tissue. To use these sequences to create controlled release, similar to the native tissue environment, biological signals are encapsulated with different polymer molecules that have different degradable rates. Such degradable rates have to fit each sequence. For instance, the first biological signal, which can be active in the early step of tissue regeneration, must be encapsulated for a higher degradable rate than a signal which is activated in the last step. The encapsulated signals have to be added into the scaffolds during the fabrication process (Fig. 7.4).

The addition of incorporated signals in scaffolds is classified into two approaches. First, the encapsulated signals are mixed with the polymer solution or melting before fabrication into the scaffolds [36]. In this process, the encapsulated signals are embedded into the polymer matrix. During this process, the biological signals can be active to induce tissue regeneration when the polymer matrix erodes or is degraded. After the polymer matrix has eroded or degraded, the encapsulated biological signals emerge from the polymer matrix. The polymers encapsulated around those signals are degraded; then, the signals can be activated to induce tissue regeneration (Fig. 7.5).

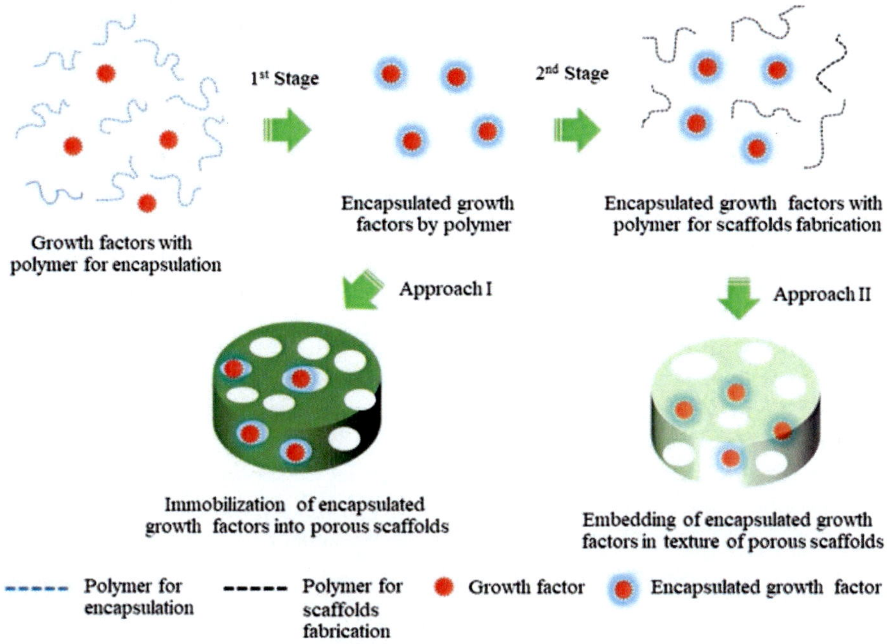


**Fig. 7.4** Scaffolds with the added encapsulated signals

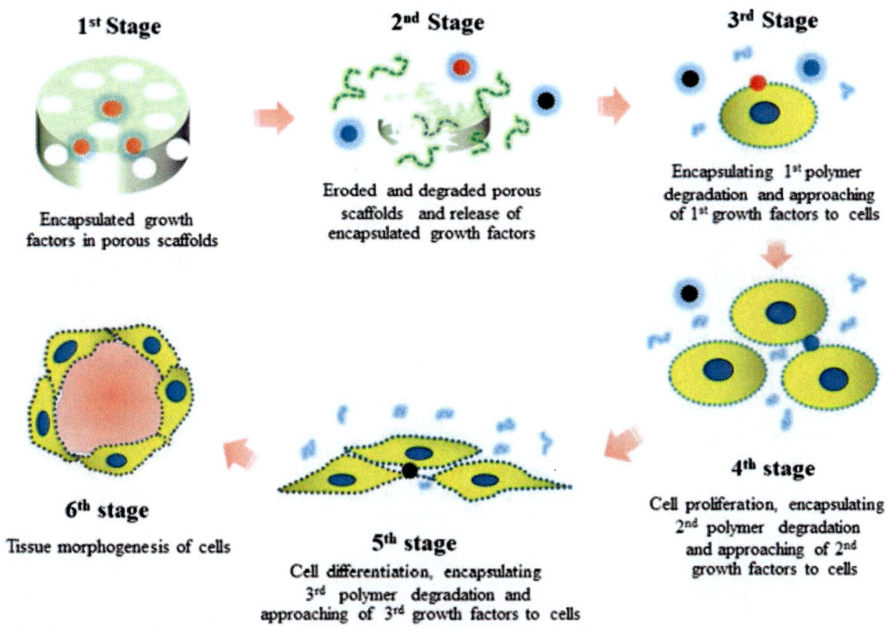


**Fig. 7.5** Sequences of activation to induce tissue regeneration from encapsulated biological signals in the polymer matrix

Second, the encapsulated biological signals are either coated or immobilized on the scaffolds [66]. During this process, the encapsulated biological signals can adhere to the surface of the scaffolds via physical interaction. The physical interactions are as follows: electrostatic forces, hydrophilic or hydrophobic interaction, and hydrogen bonding between the surface of the scaffolds and the polymers, which are encapsulated with the biological signals. When the coated or immobilized scaffolds are used for tissue regeneration, the polymers erode or degrade before emerging to activate the cells for regulation into the proposed tissue. The sequences of activation for tissue regeneration, from coated or immobilized scaffolds, by encapsulated biological signals are shown in Fig. 7.6.

As detailed in this chapter, the mimicking of biological function in scaffolds is focused with based three parts: (1) biological molecules of ECMs and growth factors in the microenvironment, (2) base polymers for scaffolds, (3) approaches to mimic biological function in scaffolds. Some instances were presented as a guidance to create performance scaffolds, based on mimicking of biological functions. This mimicking of biological function and structure in scaffolds is used in maxillofacial and articular cartilage surgery: as laid out in the next chapters.

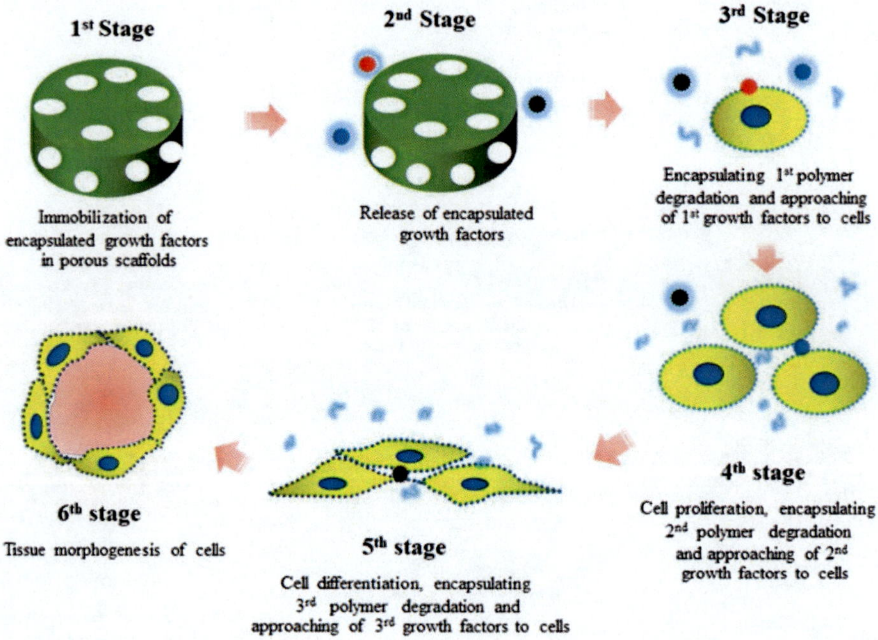


**Fig. 7.6** Sequences of activation for tissue regeneration, from coated or immobilized scaffolds, by encapsulated biological signals

## References


1. Kim, T.G., Shin, H., Lim, D.W.: Biomimetic scaffolds for tissue engineering. Adv. Funt. Mater. **22**, 2446–2468 (2012)
2. Calejo, I., Costa-Almeida, R., Reis, R.L., Gomes, M.E.: Enthesis tissue engineering: biological requirements meet at the interface. Tissue Eng. Pt B-Rev. **25**, 330–356 (2019)
3. Umuhoza, D., Yang, F., Long, D., Hao, Z., Dai, J., Zhao, A.: Strategies for tuning the biodegradation of silk fibroin-based materials for tissue engineering applications. ACS Biomater. Sci. Eng. **6**(3), 1290–1310 (2020)
4. Wu, Y.H.A., Chiu, Y.C., Lin, Y.H., Ho, C.C., Shie, M.Y., Chen, Y.W.: 3D-printed bioactive calcium silicate/poly-ε-caprolactone bioscaffolds modified with biomimetic extracellular matrices for bone regeneration. Int. J. Mol. Sci. **20**(4), 942 (2019)
5. Lei, B., Shin, K.H., Noh, D.Y., Jo, I.H., Koh, Y.H., Choi, W.Y., Kim, H.E.: Nanofibrous gelatin–silica hybrid scaffolds mimicking the native extracellular matrix (ECM) using thermally induced phase separation. J. Mater. Chem. **22**, 14133–14140 (2012)
6. Xu, Q., Zhang, Z., Xiao, C., He, C., Chen, X.: Injectable polypeptide hydrogel as biomimetic scaffolds with tunable bioactivity and controllable cell adhesion. Biomacromol. **8**(4), 1411–1418 (2017)
7. Vikulina, A.S., Skirtach, A.G., Volodkin, D.: Hybrids of polymer multilayers, lipids, and nanoparticles: mimicking the cellular microenvironment. Langmuir **35**(26), 8565–8573 (2019)
8. Liu, H., Xu, X., Tu, Y., Chen, K., Li, S., Zhai, J., Chen, S., Rong, L., Zhou, L., Wu, W., So, K.F., Ramakrishna, S., Liumin, He, L.: Engineering microenvironment for endogenous neural regeneration after spinal cord injury by reassembling extracellular matrix. ACS Appl. Mater. Interfaces **12**(15), 17207–17219 (2020)
9. Najafi, M., Farhood, B., Mortezaee, K.: Extracellular matrix (ECM) stiffness and degradation as cancer drivers. J. Cell Biochem. **120**, 2782–2790 (2019)
10. Laronha, H., Caldeira, J.: Structure and function of human matrix metalloproteinases. Cells **9**(5), 1076 (2020)
11. Hosseini, S., Naderi-Manesh, H., Vali, H., Eslaminejad, M.B., Sayahpour, F.A., Sheibanie, S., Faghihi, S.: Contribution of osteocalcin-mimetic peptide enhances osteogenic activity and extracellular matrix mineralization of human osteoblast-like cells. Colloid Surf. B **173**, 662–671 (2019)
12. Zhang, W., Yu, X., Li, Y., Su, Z., Jandt, K.D., Wei, G.: Protein-mimetic peptide nanofibers: Motif design, self-assembly synthesis, and sequence-specific biomedical applications. Prog. Polym. Sci. **80**, 94–124 (2018)
13. Huettner, N., Dargaville, T.R., Forget, A.: Discovering cell-adhesion peptides in tissue engineering: beyond RGD. Trends Biotechnol. **36**, 372–383 (2018)
14. Duncan, H.F., Kobayashi, Y., Shimizu, E.: Growth factors and cell homing in dental tissue regeneration. Curr. Oral Health Rep. **5**, 276–285 (2018)
15. Klimek, K., Ginalska, G.: Proteins and peptides as important modifiers of the polymer scaffolds for tissue engineering applications—a review polymers (Basel) **12**(4), 844 (2020)
16. Patel, R., Monticone, D., Lu, M., Grøndahl, L., Huang, H.: Hydrolytic degradation of porous poly(hydroxybutyrate-co-hydroxyvalerate) scaffolds manufactured using selective laser sintering. Polym. Degrad. Stabil. **187**, 109545 (2021)
17. van Haaften, E.E., Duijvelshoff, R., Ippel, B.D., Söntjens, S.H.M., van Houtem, M.H.C.J., Janssen, H.M., Smits, A.I.P.M., Kurniawan, N.A., Dankers, P.Y.W., Bouten, C.V.C.: The degradation and performance of electrospun supramolecular vascular scaffolds examined upon in vitro enzymatic exposure. Acta. Biomater. **92**, 48–59 (2019)
18. Zhang, L., Liu, X., Li, G., Wang, P., Yang, Y.: Tailoring degradation rates of silk fibroin scaffolds for tissue engineering. J. Biomed. Mater. Res. A **7**, 104–113 (2019)
19. Kim, K., Yu, M., Zong, X., Chiu, J., Fang, D., Seo, Y.S., Hsiao, B.S., Chu, B., Hadjiargyrou, M.: Control of degradation rate and hydrophilicity in electrospun non-woven poly(d, l-lactide) nanofiber scaffolds for biomedical applications. Biomaterials **24**, 4977–4985 (2003)

20. Sadeghi, A., Moztarzadeh, F., Mohandesi, J.A.: Investigating the effect of chitosan on hydrophilicity and bioactivity of conductive electrospun composite scaffold for neural tissue engineering. Inter. J. Biol. Macromol. **121**, 625–632 (2019)
21. Bartnikowski, M., Dargaville, T.R., Ivanovski, S., Hutmacher, D.W.: Degradation mechanisms of polycaprolactone in the context of chemistry, geometry and environment. Prog. Polym. Sci. **96**, 1–20 (2019)
22. Chiellini, E., Corti, A., Sarto, G.D., D'Antone, S.: Oxo-biodegradable polymers—effect of hydrolysis degree on biodegradation behaviour of poly(vinyl alcohol). Polym Degrad. Stabil **91**, 3397–3406 (2006).
23. Zanela, J., Casagrande, M., Reis, M.O., Victória, M., Grossmann, E., Yamashita, F.: Biodegradable sheets of starch/Polyvinyl Alcohol (PVA): effects of PVA molecular weight and hydrolysis degree. Waste Biomass Valor. **10**, 319–326 (2019)
24. Guo, X., Carter, M.C.D., Appadoo, V., Lynn, D.M.: Tunable and selective degradation of amine-reactive multilayers in acidic media. Biomacromol. **20**(9), 3464–3474 (2019)
25. Urbánek, T., Jäger, E., Jäger, A., Martin, H.M.: Selectively biodegradable polyesters: nature-inspired construction materials for future biomedical applications. Polymers **11**(6), 1061 (2019)
26. Swainson, S.M.E., Taresco, V., Pearce, A.K., Clapp, L.H., Ager, B., McAllister, M., Bosquillon, C., Garnett, M.C.: Exploring the enzymatic degradation of poly(glycerol adipate). Eur. J. Pharm. Biopharma **142**, 377–386 (2019)
27. Chen, K., Liao, S., Guo, S., Zheng, X., Wang, B., Duan, Z., Zhang, H., Gong, Q., Luo, K.: Multistimuli-responsive PEGylated polymeric bioconjugate-based nano-aggregate for cancer therapy. Chem. Eng. J. **391**, 123543 (2020)
28. Patel, J.M., Saleh, K.S., Burdick, J.A., Mauck, R.L.: Bioactive factors for cartilage repair and regeneration: Improving delivery, retention, and activity. Acta Biomater. **93**, 222–238 (2019)
29. Chun, S.Y., Lim, J.O., Lee, E.H., Han, M.H., Ha, Y.S., Lee, J.N., Kim, B.S., Park, M.J., Yeo, M.G., Jung, B., Kwon, T.G.: Preparation and characterization of human adipose tissue-derived extracellular matrix, growth factors, and stem c: a concise review. Tissue Eng. Regen. Med. **16**, 385–393 (2019)
30. Jin, Q., Liu, G., Li, S., Yuan, H., Yun, Z., Zhang, W., Zhang, S., Dai, Y., Ma, Y.: Decellularized breast matrix as bioactive microenvironment for in vitro three-dimensional cancer culture. J. Cell Physiol. **234**, 3425–3435 (2019)
31. Kim, H., Bae, C., Kook, Y.M., Koh, W.G., Lee, M., Park, M.H.: Mesenchymal stem cell 3D encapsulation technologies for biomimetic microenvironment in tissue regeneration. Stem Cell Res. Ther. **10**, 51 (2019)
32. Sangkert, S., Kamonmattayakul, S., Chai, W.L., Meesane, J.: Modified silk and chitosans with collagen assembly for osteopoeosis. Bioinspirated Biomimet. Nanobiomat. **5**(1), 1–11 (2016)
33. Jaipaew, J., Wangkulangkul, P., Meesane, J., Raungrut, P., Puttawibul, P.: Mimicked cartilage scaffolds of silk fibroin/hyaluronic acid with stem cells for osteoarthritis surgery: morphological, mechanical, and physical clues. Mat. Sci. Eng. C-Mater. **64**, 173–182 (2016)
34. Dinoro, J., Maher, M., Talebian, S., Jafarkhani, M., Mehrali, M., Orive, G., Foroughi, J., Lord, M.S., Dolatshahi-Pirouz, A.: Sulfated polysaccharide-based scaffolds for orthopaedic tissue engineering. Biomaterials **214**, 119214 (2019)
35. Lia, M., Zhang, A., Li, J., Zhou, J., Zheng, Y., Zhang, C., Xi, D., Mao, H., Zhao, J.: Osteoblast/fibroblast coculture derived bioactive ECM with unique matrisome profile facilitates bone regeneration. Bioact. Mater. **5**, 938–948 (2020)
36. Chen, L., Liu, J., Guan, M., Zhou, T., Duan, X., Zhou, X.Z.: Growth factor and its polymer scaffold-based delivery system for cartilage tissue engineering. Int. J. Nanomedicine **15**, 6097–6111 (2020)
37. Rajpar, I., Barrett, J.G.: Optimizing growth factor induction of tenogenesis in three-dimensional culture of mesenchymal stem cells. J. Tissue Eng. **10**, 1–9 (2019)
38. Fahimipour, F., Dashtimoghadama, E., Hasani-Sadrabadi, M.M., Vargas, J., Vashaee, D., Lobner, D.C., Ghasemzadeh, B., Tayebi, L.: Enhancing cell seeding and osteogenesis of MSCs on 3D printed scaffolds through injectable BMP2 immobilized ECM-Mimetic gel. Dent. Mater. **35**, 990–1006 (2019)

39. Kenar, H., Ozdoganad, C.Y., Dumlu, C., Doger, E., Kose, G.T., Hasirci, V.: Microfibrous scaffolds from poly(l-lactide-co-ε-caprolactone) blended with xeno-free collagen/hyaluronic acid for improvement of vascularization in tissue engineering applications. Mat. Sci. Eng. C-Mater **97**, 31–44 (2019)
40. Puttawibul, P., Benjakul, S., Meesane, J.: An in situ hydrogel of mimicked self-assembly type I collagen from shark skin (brownbanded bamboo shark, Chiloscyllium punctatum)/methylcellulose for central nerve system regeneration: preparation and characterization. J. Biomim. Biomater. Biomed. Eng. **14**, 14–29 (2015)
41. Moxona, S.R., Corbett, N.J., Fisher, K., Potjewyd, G., Domingos, M., Hooper, N.M.: Blended alginate/collagen hydrogels promote neurogenesis and neuronal maturation. Mat. Sci. Eng. C-Mater. **104**, 109904 (2019)
42. Liu, C., Jin, Z., Ge, X., Zhang, Y., Xu, H.: Decellularized annulus fibrosus matrix/chitosan hybrid hydrogels with basic fibroblast growth factor for annulus fibrosus tissue engineering. Tissue Eng. A **25**, 1605–1613 (2019)
43. Hoshino, Y., Lee, H., Miura, Y.: Interaction between synthetic particles and biomacromolecules: fundamental study of nonspecific interaction and design of nanoparticles that recognize target molecules. Polym. J. **46**, 537–545 (2014)
44. Auría-Soro, C., Nesma, T., Juanes-Velasco, P., Landeira-Viñuela, A., Fidalgo-Gomez, H., Acebes-Fernandez, V., Gongora, R., Parra, M.J.A., Manzano-Roman, R., Fuentes, M.: Interactions of nanoparticles and biosystems: microenvironment of nanoparticles and biomolecules in nanomedicine. Nanomaterials (Basel) **9**(10), 1365 (2019)
45. Ata, S., Rasool, A., Islam, A., Bibi, I., Rizwan, M., Azeem, K.M., Qureshi, A.R., Iqbal, M.: Loading of Cefixime to pH sensitive chitosan based hydrogel and investigation of controlled release kinetics. Int. J. Biol. Macromol. **155**, 1236–1244 (2020)
46. Subbiah, R., Guldberg, R.E.: Materials science and design principles of growth factor delivery systems in tissue engineering and regenerative medicine. Adv. Health Mater. **8**, 1801000 (2019)
47. Niu, Y., Li, Q., Ding, Y., Dong, L., Wang, C.: Engineered delivery strategies for enhanced control of growth factor activities in wound healing. Adv. Drug Deliver. Rev. **146**, 190–208 (2019)
48. Stupp, S.I., Zha, R.H., Palmer, L.C., Cui, H., Bitton, R.: Self-assembly of biomolecular soft matter. Faraday Discuss **166**, 9–30 (2013)
49. Putttawibul, P., Benjakul, S., Meesane, J.: Freeze-Thawed hybridized preparation with biomimetic self-assembly for a polyvinyl alcohol/collagen hydrogel created for meniscus tissue engineering. J. Biomim. Biomater Biomed. Eng. **21**, 17–33 (2014)
50. Danesin, R., Brun, P., MartinaRoso, M., Delaunay, F., Samouillan, V., Brunelli, K., Iucci, G., Ghezzo, F., Modesti, M., Castagliuolo, I., Dettin, M.: Self-assembling peptide-enriched electrospun polycaprolactone scaffolds promote the h-osteoblast adhesion and modulate differentiation-associated gene expression. Bone **51**, 851–859 (2012)
51. Chen, L., Mou, S., Li, F., Zeng, Y., Sun, Y., Horch, R.E., Wei, W., Wang, Z., Jiaming, S.J.: Self-assembled human adipose-derived stem cell-derived extracellular vesicle-functionalized biotin-doped polypyrrole titanium with long-term stability and potential osteoinductive ability. ACS Appl. Mater. Interfaces **11**(49), 46183–46196 (2019)
52. Yan, M., Jiang, X., Wang, G., Wang, A., Wang, X., Wang, X., Zhao, X., Xu, H., An, X., Li, Y.: Preparation of self-assembled collagen fibrillar gel from tilapia skin and its formation in presence of acidic polysaccharides. Carbohyd. Polym. **233**, 115831 (2020)
53. Li, X., Wang, L., Li, L., Luo, Z., Yan, S., Zhang, Q., You, R.: Water-stable silk fibroin nerve conduits with tunable degradation prepared by a mild freezing-induced assembly. Polym. Degrad. Stabil. **164**, 61–68 (2019)
54. Nerantzaki, M., Loth, C., Lutz, J.F.: Chemical conjugation of nucleic acid aptamers and synthetic polymers. Polym. Chem. **2**, 3498–3509 (2021)
55. Hajimiri, M., Shahverdi, S., Kamalinia, G., Dinarvand, R.: Growth factor conjugation: Strategics and applications. J. Biomed. Mater. Res. A **103**, 819–838 (2015)
56. Thakar, H., Sebastian, S.M., Mandal, S., Pople, A., Agarwal, G., Srivastava, A.: Biomolecule-conjugated macroporous hydrogels for biomedical applications. ACS Biomater. Sci. Eng. **5**(12), 6320–6341 (2019)

57. Duan, Y., Yu, S., Xu, P., Wang, X., Feng, X., Mao, Z., Gao, C.: Co-immobilization of CD133 antibodies, vascular endothelial growth factors, and REDV peptide promotes capture, proliferation, and differentiation of endothelial progenitor cells. Acta Biomater. **96**, 137–148 (2019)
58. Arisaka, Y., Yui, N.: Engineering molecularly mobile polyrotaxane surfaces with heparin-binding EGF-like growth factors for improving hepatocyte functions. J. Biomed. Mater. Res. A **107**, 1080–1085 (2019)
59. Amani, H., Arzaghi, H., Bayandori, M., Dezfuli, A.S., Pazoki-Toroudi, H., Shafiee, A., Moradi, L.: Controlling cell behavior through the design of biomaterial surfaces: a focus on surface modification techniques. Adv. Mater. Interfaces **6**, 1900572 (2019)
60. Pompe, T., Salchert, K., Alberti, K., Zandstra, P., Werner, C.: Immobilization of growth factors on solid supports for the modulation of stem cell fate. Nat Protoc **5**, 1042–1050 (2010)
61. Hao, D., Fan, Y., Xiao, W., Liu, R., Pivetti, C., Walim, T., Guo, F., Zhang, X., Farmer, D.L., Wang, F., Panitch, A., Slam, K.S., Wang, A.: Rapid endothelialization of small diameter vascular grafts by a bioactive integrin-binding ligand specifically targeting endothelial progenitor cells and endothelial cells. Acta Biomater. **108**, 178–193 (2020)
62. Alipour, M., Baneshi, M., Hosseinkhani, S., Mahmoudi, R., Arabzadeh, A.J., Akrami, M., Mehrzad, J., Bardania, H.: Recent progress in biomedical applications of RGD-based ligand: From precise cancer theranostics to biomaterial engineering: A systematic review. J. Biomed. Mater. Res. A **108**, 839–850 (2020)
63. Tallawi, M., Rosellini, E., Barbani, N., Cascone, M.G., Rai, R., Saint-Pierre, G., Boccaccini, A.R.: Strategies for the chemical and biological functionalization of scaffolds for cardiac tissue engineering: a review. J. R Soc. Interface **12**(108), 20150254 (2015)
64. McLean, K.M., Johnson, G., Chatelier, R.C., Beumer, G.J., Steele, J.G., Griesser, H.J.: Method of immobilization of carboxymethyl-dextran affects resistance to tissue and cell colonization. Colloid Surf. B **18**, 221–234 (2000)
65. Saik, J.E., Gould, D.J., Watkins, E.M., Dickinson, M.E., West, J.L.: Covalently immobilized platelet-derived growth factor-BB promotes angiogenesis in biomimetic poly(ethylene glycol) hydrogels. Acta Biomater. **7**, 133–143 (2011)
66. Witte, T.M., Wagner, A.M., Fratila-Apachitei, L.E., Zadpoor, A.A., Peppas, N.A.: Immobilization of nanocarriers within a porous chitosan scaffold for the sustained delivery of growth factors in bone tissue engineering applications. J. Biomed. Mater. Res. A **108**, 1122–1135 (2020)
67. Ma, Z., He, W., Yong, T., Ramakrishna, S.: Grafting of gelatin on electrospun poly(caprolactone) nanofibers to improve endothelial cell spreading and proliferation and to control cell orientation. Tissue Eng. **11**, 1149–1158 (2005)
68. Schmedlen, R.H., Masters, K.S., West, J.L.: Photocrosslinkable polyvinyl alcohol hydrogels that can be modified with cell adhesion peptides for use in tissue engineering. Biomat. **23**, 4325–4332 (2002)
69. Rafat, M., Li, F., Fagerholm, P., Lagali, N.S., Watsky, M.A., Munger, R., Matsuura, T., Griffith, M.: PEG-stabilized carbodiimide crosslinked collagen–chitosan hydrogels for corneal tissue engineering. Biomaterials **29**, 3960–3972 (2008)
70. Vijayan, A., Sabareeswaran, A., Vinod Kumar, G.S.: PEG grafted chitosan scaffold for dual growth factor delivery for enhanced wound healing. Sci. Rep. **9**, 19165 (2019)
71. Pethő, L., Kasza, G., Lajkó, E., Láng, O., Kőhidai, L., Iván, B., Mező, G.: Amphiphilic drug–peptide–polymer conjugates based on poly(ethylene glycol) and hyperbranched polyglycerol for epidermal growth factor receptor targeting: the effect of conjugate aggregation on in vitro activity. Soft Matter. **16**, 5759–5769 (2020)
72. Wang, J., Su, G., Yin, X., Luo, J., Gu, R., Wang, S., Feng, J., Chen, B.: Non-small cell lung cancer-targeted, redox-sensitive lipid-polymer hybrid nanoparticles for the delivery of a second-generation irreversible epidermal growth factor inhibitor—Afatinib: In vitro and in vivo evaluation. Biomed. Pharm. **120**, 109493 (2019)
73. Zhang, W., Shi, W., Wu, S., Kuss, M., Jiang, X., Untrauer, J.B., Reid, S.P., Duan, B.: 3D printed composite scaffolds with dual small molecule delivery for mandibular bone regeneration. Biofabric. **12**, 035020 (2020)

74. Kim, T.G., Shin, H., Lim, D.W.: Biomimetic scaffolds for tissue engineering. Adv. Funct. Mater. **22**, 2446–2468 (2012)
75. Richbourg, N.R., Peppas, N.A., Sikavitsas, V.I.: Tuning the biomimetic behavior of scaffolds for regenerative medicine through surface modifications. J. Tissue Eng. Regen. M **13**, 1275–1293 (2019)
76. Simon, Y., Patel, Z.S., Kretlow, J.D., Murphy, M.B., Mountziaris, P.M., Baggett, L.S., Ueda, H., Tabata, Y., Jansen, J.A., Wong, M., Mikos, A.G.: Dose effect of dual delivery of vascular endothelial growth factor and bone morphogenetic protein-2 on bone regeneration in a rat critical-size defect model. Tissue Eng. A **15**, 2347–2362 (2009)

# Chapter 8
# Mimicked 3D Scaffolds for Maxillofacial Surgery

For maxillofacial surgery, scaffolds are used as performance biomaterials for enhancing new tissue formation [1, 2]. In the view of tissue engineering, scaffolds have the main function as the structure supporting cell behaviors [3]. On the other hand, from the view of maxillofacial surgery, scaffolds show the function as supporting material to complete the anatomy of organs which have tissue defects [4]. For instance, scaffolds have the function to act as degradable materials, which assist to complete the function and contour shape after surgery, without dysfunction of the organ or residual material debris at the defect site [5]. Therefore, to create scaffolds for maxillofacial surgery, one needs comprehension regarding both tissue engineering and surgical approach.

Mimicking is the approach which is used to design the scaffolds for tissue engineering [5]. Scaffolds which are created, with based mimicking, demonstrate both structure and function similar to the natural extracellular matrix (ECM) of tissue [6]. Material selection and fabrication is the key to mimic scaffolds, which are similar to natural ECMs [7]. For example, collagen type I, which is the main component in tissue, is often selected as the base material to mimic scaffolds [8]. Self-assembly is the fabrication technique that is used to mimic scaffolds similar to the fibril structure of natural ECMs [9]. Micro- and nano-structures coupled with physical and biological function are often used to mimic scaffolds [10].

In contrast, for maxillofacial surgery, a macro-structure fitting to the anatomy of an organ is used to mimic scaffolds [4]. Suitable function for practical operation is focused on, so as to combine with the mimicked macro-structure of said scaffolds. For instance, a mimicked, macro-structure of the scaffold is created with a personal design, based on three dimensional (3D) printing [11]. Scaffolds are fabricated into the geometry that is similar to the anatomy of each patient. For some cases, these mimicked scaffolds need the ability of molding for fitting in with the tissue defect. This is the requirement for practical operation [12, 13].

In this chapter, mimicked scaffolds, based on the view of tissue engineering and surgery of maxillofacial surgery, are explained, along with examples of mimicked



J. Meesane, *Mimicked Tissue Engineering Scaffolds for Maxillofacial and Articular Cartilage Surgery*, Engineering Materials,
https://doi.org/10.1007/978-981-19-7830-2_8

scaffolds for maxillofacial surgery are demonstrated. Furthermore, some ideas to develop mimicked scaffolds are focused on.

## 8.1 3D Scaffolds for Maxillofacial Surgery

Many patients suffer from the problem associated with maxillofacial defects. Maxillofacial defects normally arise from either trauma or diseases. In some cases, the patient requires surgery for bone replacement at the defect area. The areas of maxillofacial surgery, which are explained in this chapter, focus on the defect of the mandible and maxilla: bone resorption and alveolar cleft lip and palate (Fig. 8.1).

Generally, maxillofacial defects are critical problems for many patients, and in severe cases, patients need an operation involving bone grafting or substitution with performance biomaterials (Fig. 8.2a and b) [14]. In the case of auto-bone grafting, the bone is removed from a healthy area of the patient's own body. Then, the healthy bone is placed at the bone defect [15]; however, this technique then creates a new defect at the site of the removed bone. Therefore, to solve this problem, performance biomaterials, metals, ceramic, non-degradable polymers, and non-degradable composites are designed into specific geometries to be placed at the defect site (Fig. 8.2b) [14, 16]. However, in the long term, biomaterials can produce debris, which the body identifies as foreign bodies and are then subject to rejection by the body [17]. Furthermore, when fixation at the defect site is not sufficiently stable, the patients need to have a second operation.

To solve the problems of bone grafting as well as biomaterial implantation, 3D scaffolds have been selected as performance biomaterials to induce new tissue formation at the defect site [18]. To meet the requirements of biocompatibility and non-toxicity, 3D scaffolds are attractive as a substitute material at a defect site [19]. The main function of 3D scaffolds is to maintain the contour shape of the defect until new bone formation is complete at said defect site [20]. Furthermore, the scaffolds must degrade within a specific period of time for complete new bone formation [21].

Some scaffolds are developed for further function to induce new tissue formation or to prevent infectious at the defect site [22]. For example, some bioactive compounds that have the function to induce bone formation are added into the scaffolds [23]. Some scaffolds are incorporated with bioactive molecules or drugs, which act as local delivery system, to suppress tumor recurrence after surgery [24, 25]. Additionally, in some cases, scaffolds are also mixed with antimicrobial drugs or molecules to prevent infections during new tissue formation [26, 27].

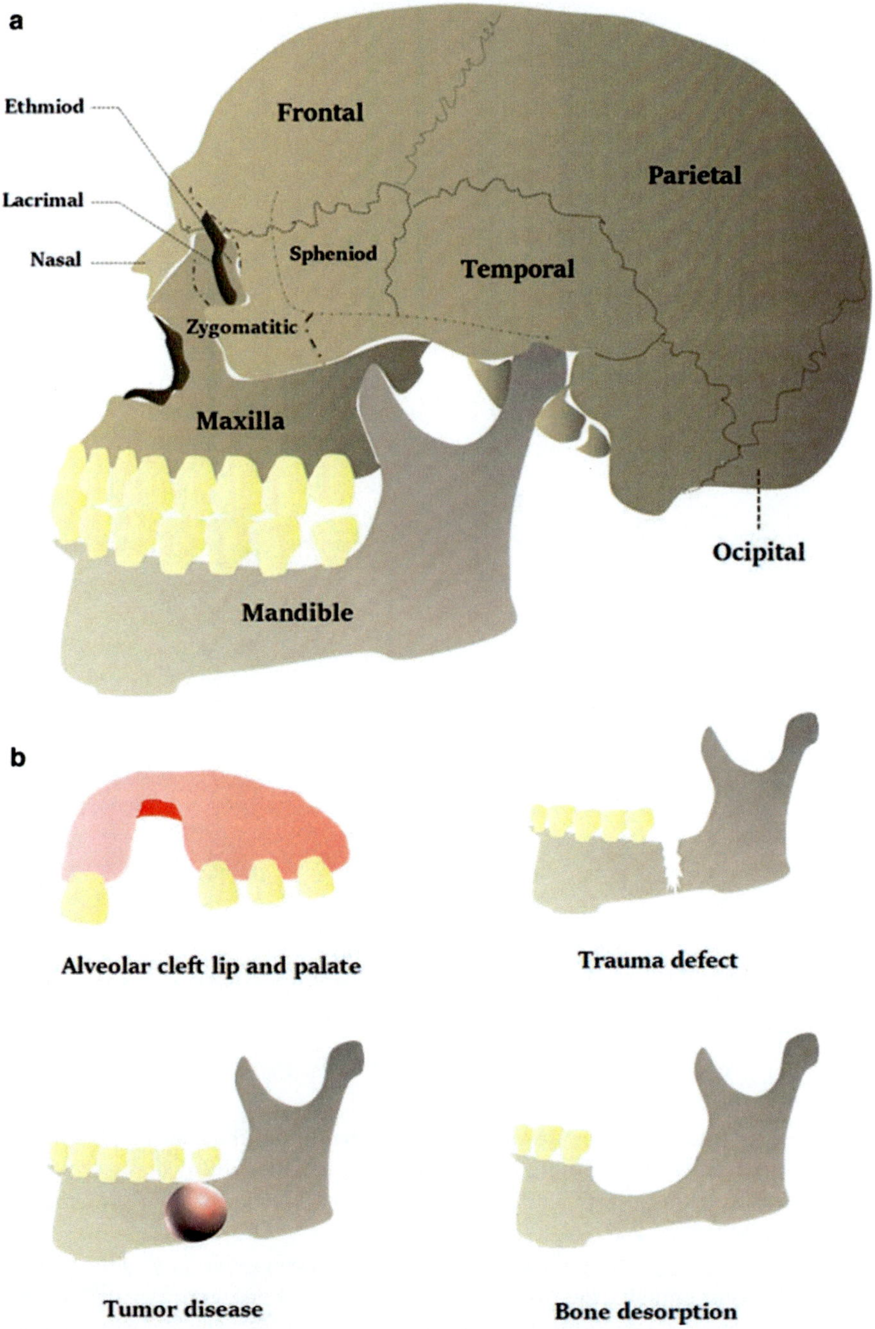


**Fig. 8.1** **a** Anatomy at maxillofacial area, **b** Example of defect within the maxillofacial area

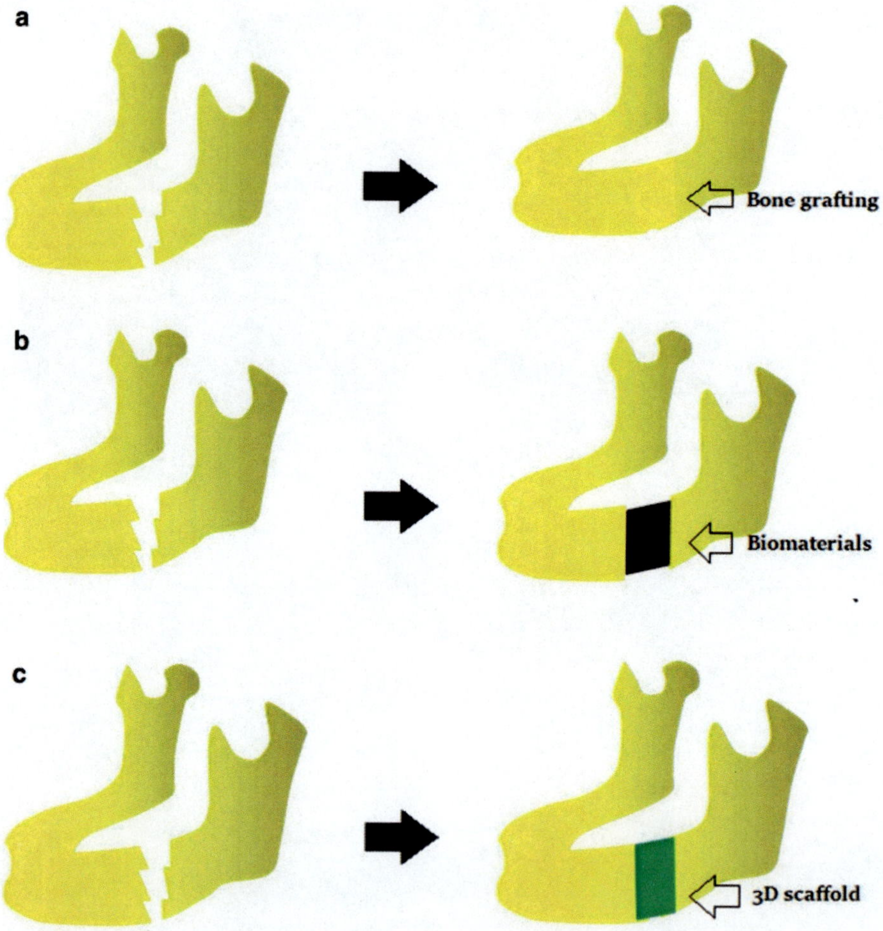


**Fig. 8.2** **a** Surgery of trauma mandible defect with bone grafting, **b** biomaterials, and **c** 3D scaffolds

### *8.1.1 3D Scaffolds for Bone Resorption at Mandible*

For bone resorption, when the patients lose their teeth, the bone at the lost area is desorbed; this leads to bone volume decreasing. Elderly patients, especially, have critical problems from tissue degenerative diseases, as their bone resorption dramatically leads to malformation of the mandible (Fig. 8.3).

Autogenous bone grafting is a surgical approach to enhance volume at the resorption area of the mandible. For some cases, autogenous bone grafting is used as a total replacement at the resorption area of the mandible [28]. Alternatively, autogenous bone grafting is used as a partial replacement for enhancement of bone volume before dental implantation (Fig. 8.4). However, autogenous bone grafting presents problems at the site of the donors bone area, which come from the patient's bone. This

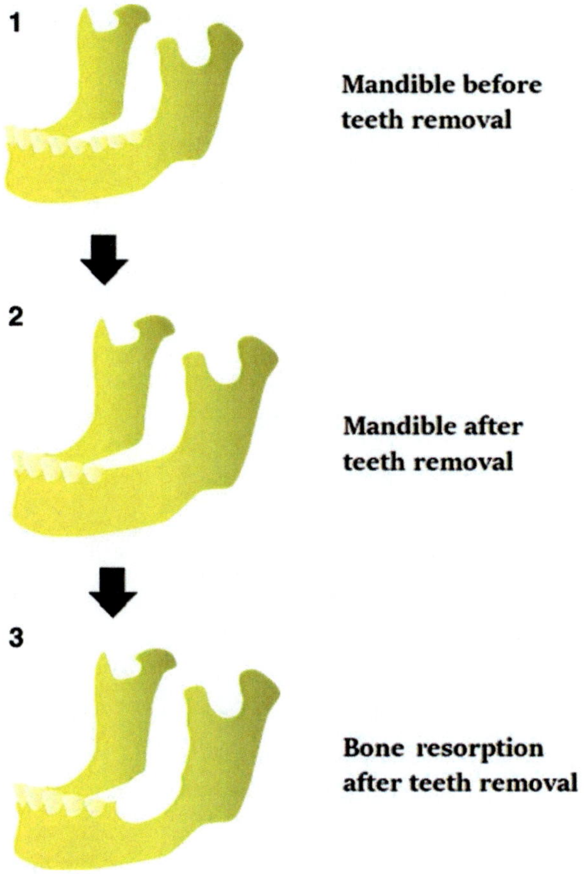


**Fig. 8.3** Progress of bone resorption at the mandibular area

causes a defect at the donor area that then requires time to heal and repair. Therefore, performance biomaterials are an attractive choice for surgery in bone resorption at the mandible.

3D scaffolds provide performance biomaterials that can enhance new tissue formation at a defect area [29]. In the case of surgery for bone resorption at the mandible, the defect area is adjusted to fit the shape of 3D scaffolds [30, 31] (Fig. 8.5). 3D scaffolds show performance to maintain the shape of the adjusted defect area until new bone has completely formed. For bone resorption, based on tissue degenerative disease, the scaffolds need added biological signals, so as to enhance their potential in the activation of new bone formation [30].

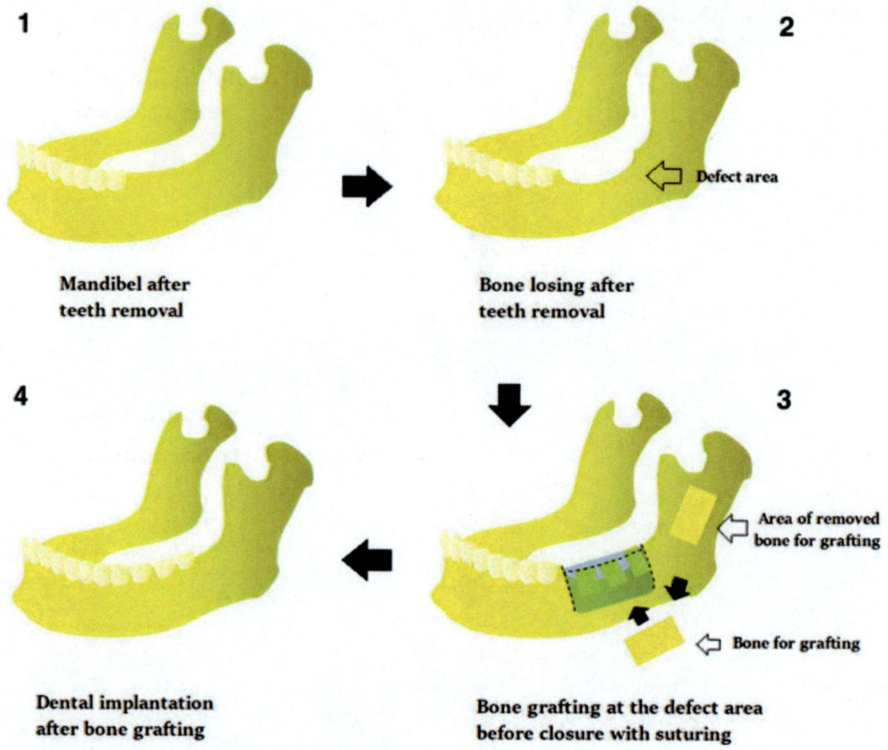


**Fig. 8.4** Bone grafting for operation of bone resorption at the mandible

## *8.1.2 3D Scaffolds for Alveolar Cleft Lip and Palate*

Alveolar cleft lip and palate is **a** congenital bone malformation that is generated from irregular development. Some studies have reported that a congenital bone and soft tissue malformation during fetal development [32]. That dysfunction causes a disordered formation of bone and soft tissue. The pathogenesis of cleft lip and palate begins with irregular tissue generation during fetal development [33]. A cleft lip and palate have incomplete bone generation around the mouth area that leads to an irregular shape of the upper lip, as illustrated in Fig. 8.6. The irregular shape of a cleft lip and palate of the divided part of the mouth is a bi-phasic defect that involves both soft and hard tissue. Patients with this type of defect often have problems with feeding and verbal communication.

Treatments for some cases require an operation [32], so surgeons need a performance method for an operation in alveolar cleft lip and palate surgeries. In some cases, where there is a complicated formation disorder, patients need a second operation [34]. Therefore, selecting the proper method that fits each type of congenital bone malformation is important before any operation. In some cases, the proper

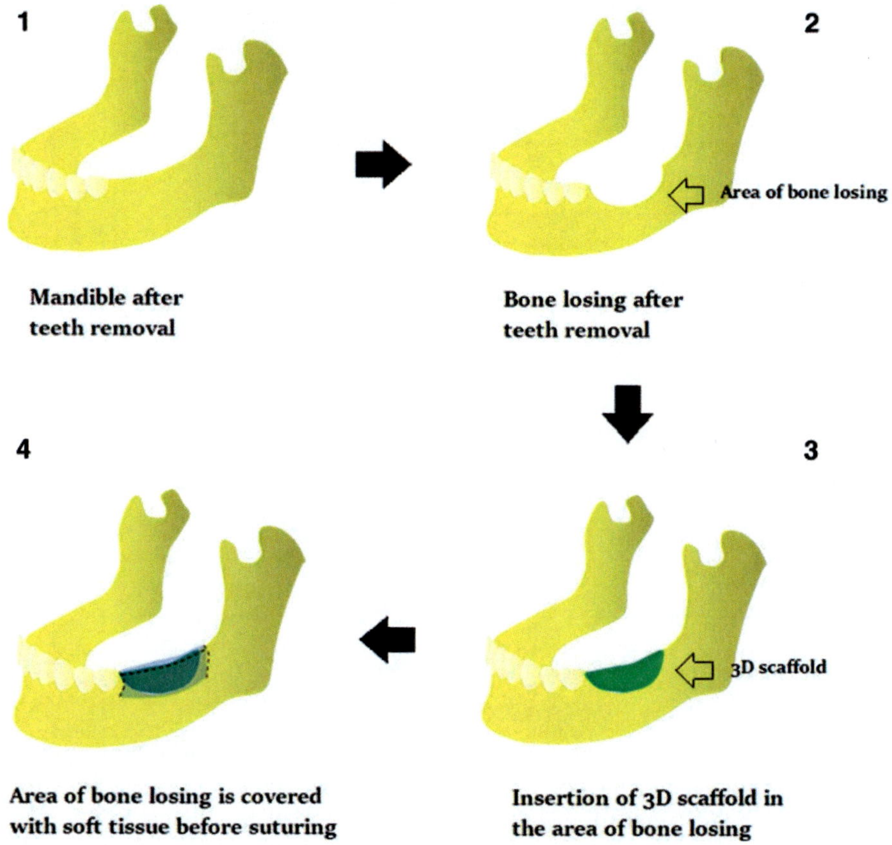


**Fig. 8.5** Adjusted defect area of alveolar bone resorption to fit biomimetic 3D scaffolds

method is to use performance biomaterials to induce and regulate bone formation [35].

For cleft lip and palate, surgeons often use bone grafting on patients (Fig. 8.7) [36]. During this approach, the patient suffers from the loss of tissue at another area. Therefore, to solve this problem, surgeons use biocompatible materials to replace the defective area [35]. A popular material used in these cases is calcium phosphate [37]. Calcium phosphate is mixed with fibrin glue to serve as a binder to maintain the contour shape at the defect site [38]. However, this approach still has only a moderate potential for new bone formation. Therefore, to enhance the potential of new bone formation, calcium phosphate adjoined with cells [39]. Some approaches have difficulty in maintaining a regular shape at the defect site. After surgeons place the aforementioned materials into the defect site, the patients will still have an irregular shape of the lip and palate, and therefore, they will require a second operation to complete the shape [40].

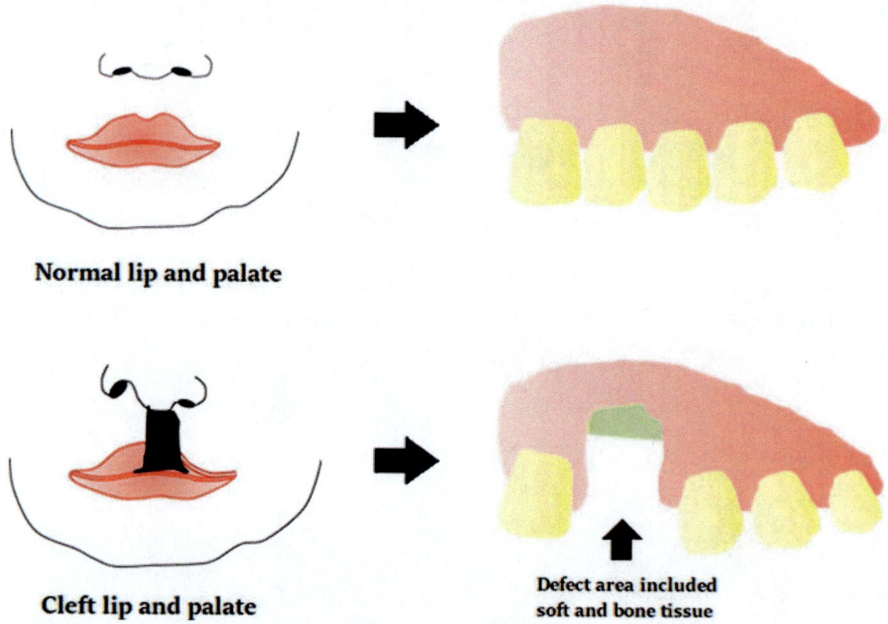


**Fig. 8.6** Cleft lip and palate and its defects

To solve the problems of bone grafting, surgeons use performance biomaterials to form a regular shape of the lip and palate [40]. Thus, creating 3D scaffolds for performance is an interesting choice for cleft lip and palate treatment [41]. The two parts of the anatomical structure at the defect area with the soft tissue include bone tissue. In the case of the defect including the total soft and bone tissue layer, flapping is used to prepare the space at the defect area for scaffold insertion. Then, the performance scaffolds are initially inserted into the defect area, before closure with suturing. Generally, for this type of operation, the scaffolds can maintain the contour shape until new bone formation is complete and replaces the defect area (Fig. 8.8).

For the case of the defect including the partial soft and bone tissue layers, the scaffold is inserted before closure with suturing (Fig. 8.9). Surgery of alveolar cleft lip and palate with 3D scaffolds has some problems regarding the invasion of soft tissue during new bone formation. This has an effect on the irregular, new tissue formation. In this case, the surgeons use a barrier membrane for protection against soft tissue invasion. The barrier membrane is used to cover the top and bottom surface of the scaffold. Furthermore, this barrier membrane also acts as the 2D scaffold for soft tissue formation. Details regarding 2D scaffolds are explained in the next chapter.

A crucial problem, which often occurs in a cleft lip and palate operation, is the incomplete sealing of the soft tissue (Fig. 8.10). This is because of the shrinkage in volume under the soft tissue area. The shrinkage in volume leads to incomplete sealing and new tissue formation. Therefore, to solve this problem, the surgeon can

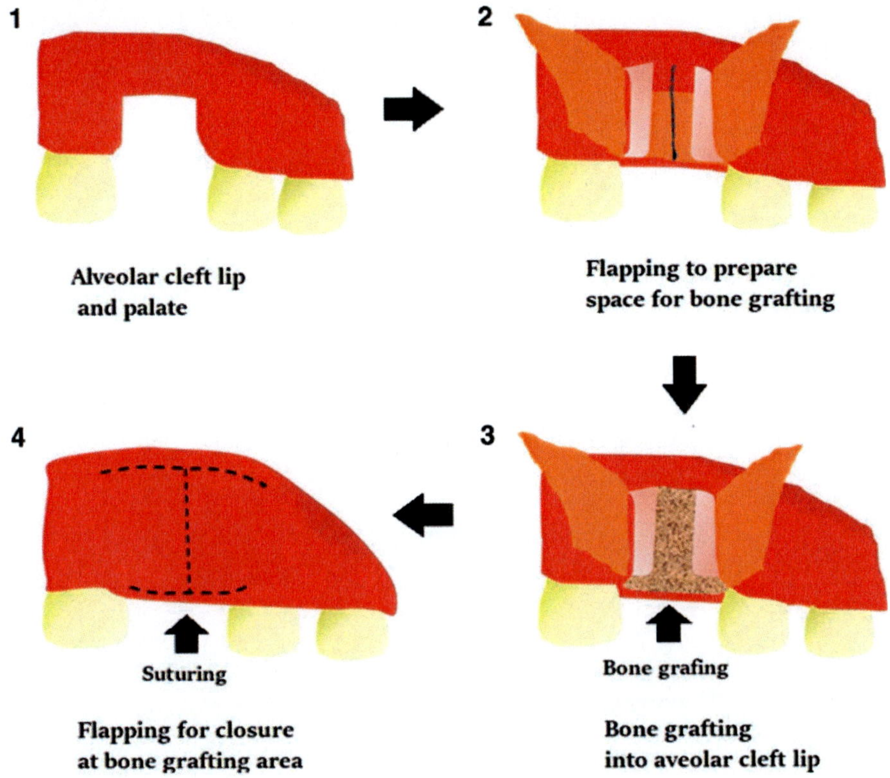


**Fig. 8.7** Surgery of alveolar cleft lip and palate with bone grafting

use hydrogel as a soft tissue expander. Hydrogel expanders expand into the shrunken volume [42, 43]. Hydrogels serve to recover and maintain the contour shape, while providing sufficient space for bone tissue regeneration. Furthermore, hydrogels also assist in the performance for new, soft tissue formation.

## 8.2 Mimicked 3D Scaffolds for Maxillofacial Defects

In mandible surgery, 3D scaffolds demonstrate suitable performance to promote new bone formation. However, in some cases, which have the large and deep damage of tissue, the scaffolds need their functional development to promote new tissue formation similar to natural tissue. From the view of tissue engineering, mimicking is the approach which is used to create scaffolds that have both structure and function similar to natural tissue. These are performance biomaterials, which have the potential to promote new tissue formation. From a surgical perspective, the mimicked scaffolds have to show suitable performance for practical operation, for instance, possibility

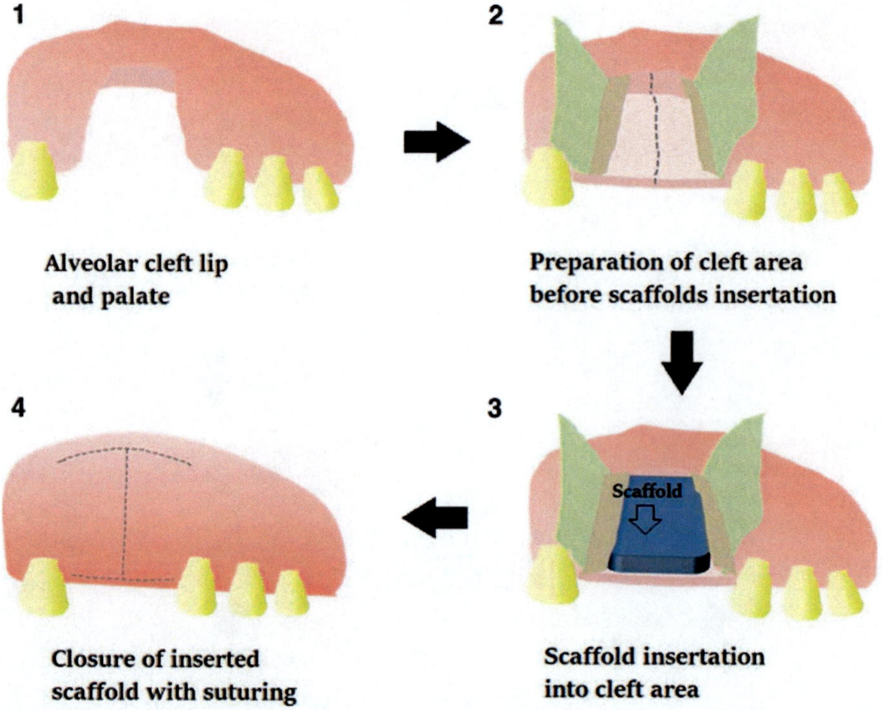


**Fig. 8.8** Surgery of alveolar cleft lip and palate, with a scaffold for the defect, including the total soft and bone tissue layer

for handling, fitting into the damage tissue area and have an ability of being molded. Therefore, to create mimicked scaffolds, one needs the optimized perspective of tissue engineering and surgery.

## *8.2.1 Mimicked Molecular Structural 3D Scaffolds*

The first mimicking for scaffolds is the structure of natural tissue. Mainly, mimicked structures focus on the extracellular matrix (ECM), which is the main component in natural tissue. Mimicked structures of ECMs are divided into three levels: (1) nano or molecular, (2) micro or morphological, (3) macro or geographical scale, as explained in previous chapters. In mimicking of ECM molecules, collagen and proteoglycan which are the main components are used in the guidance of creating scaffolds [44, 45]. Collagen is often used as a base material to mimic the structure of ECMs. At the molecule level of polypeptide, collagen peptide molecules are mimicked with the process of protein engineering [46]. Engineered molecules show a bioactive amino acid sequence similar to collagen peptide [46–48]. Furthermore,

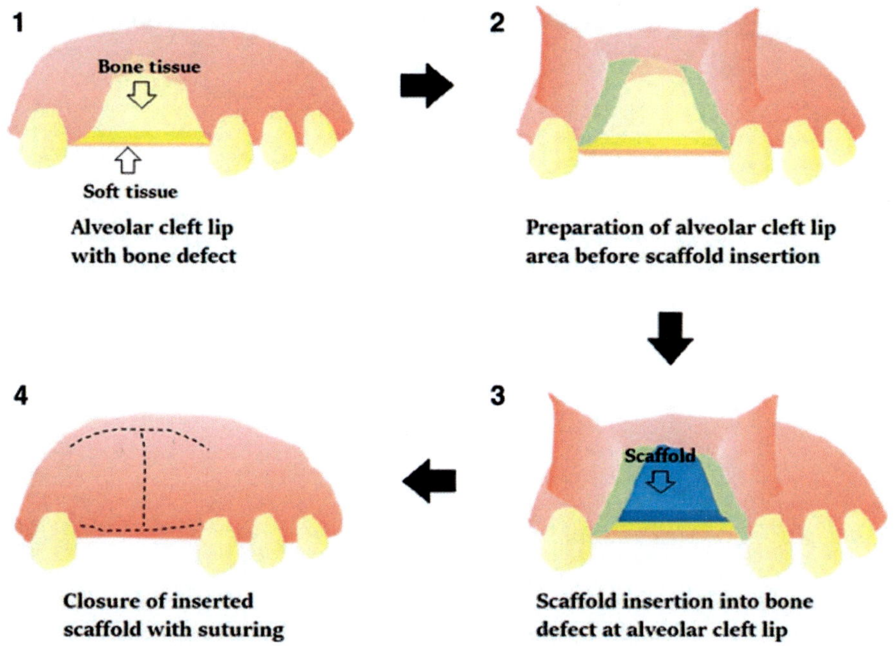


**Fig. 8.9** Surgery of alveolar cleft lip and palate, with a scaffold for the defect, including the partial soft and bone tissue layer

these molecule demonstrates self-organization also similar to natural collagen. The self-organization of peptide has an effect on promotion of cell behaviors, which in turn enhances new tissue formation [47]. For other molecular structural mimicking, some natural proteins, which have cell binding sites similar to collagen or proteoglycan in ECMs, are used to mimic scaffolds [49]. For instance, silk fibroin is often selected to mimic the structure of ECMs [11, 50]. It shows a self-organization similar to natural collagen [51]. Gelatin, that displays an amino acid sequence similar to collagen, has also been chosen as a base material to mimic the structure of ECMs [52].

In the perspective of mandible surgery, the mimicked molecular structural scaffolds need suitable performance for operation. The first requirement is the possible handling of the scaffold during an operation. For this requirement, some scaffolds have to have improved mechanical properties, which have suitable performance for handle during an operation [53]. For instance, silk fibroin must transform the molecular organization of random coils to beta sheet, which enhances the mechanical strength of the scaffold [54], while gelatin has to crosslink to increase the mechanical stability [55]. Sufficient mechanical strength and stability assist to avoid any tearing during the operation. The second requirement is the fitting to the damaged tissue area. For example, some scaffolds are mixed with antimicrobial drugs to protect against infectious at the damage tissue area [56]. For tumor diseases, some scaffolds have added bioactive molecules as a local releasing system to suppress recurrence

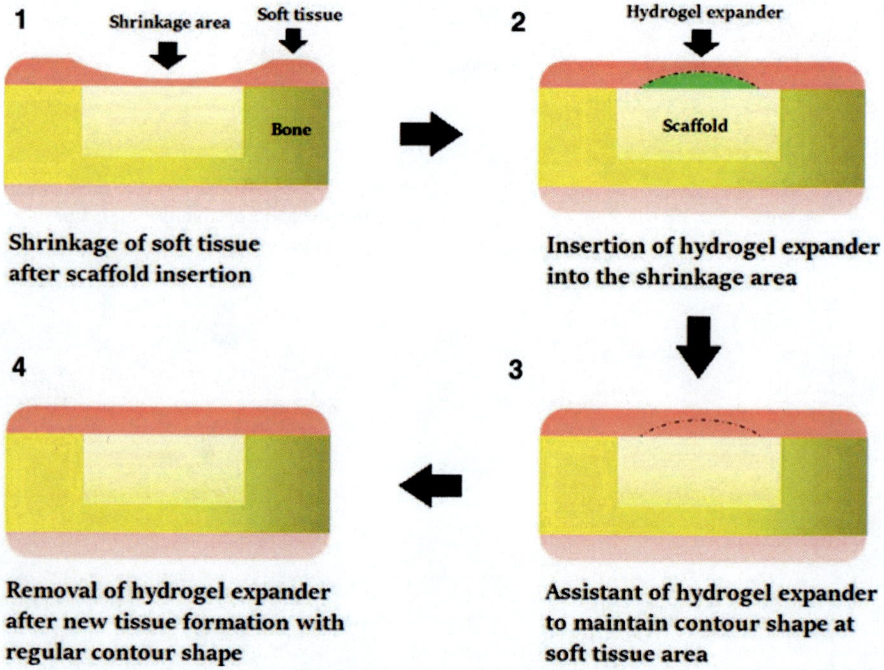


**Fig. 8.10** Operation of soft tissue shrinkage, a non-injectable hydrogel as a mucosa tissue expander

[25]. The third requirement is the ability for molding, as each, damaged tissue area has a different shape. Some scaffolds are modified for flexibility; these scaffolds show the performance for molding and suturing during surgery [57].

### *8.2.2 Mimicked Morphologically Structural 3D Scaffolds*

The second mimicking of morphological ECMs is used to create scaffolds that are suitable to promote cell behaviors, for enhancement of new tissue formations [58]. Porous and fiber structures are the main morphology which has an effect on increasing of cell behaviors: adhesion, proliferation, and migration [59, 60].

For porous structures of ECMs, it has to show interconnecting with a suitable size for cell behaviors [61, 62]. Furthermore, it has to promote nutrient diffusion, so as to assist enhancement of cell behaviors [63]. To mimic pore size and a porous structure, similar to ECMs, some scaffolds are fabricated with: practical, advanced, or integrated processing. For instance, scaffolds are simply fabricated with particle leaching to mimic the main pores, which have loose connecting as ECMs [64]. These scaffolds then continue freeze drying to mimic pore sizes similar to ECMs. Other scaffolds are fabricated from leaching of particle compounded polymers [62]. For this fabrication, the polymers are compounded with the particles by a mixer machine to break the particles into a non-homogeneous texture [58]. The compounded polymers are then processed into 3D specimens, before particle leaching, to obtain non-homogeneous porous scaffolds, which are similar to the morphology of ECMs.

In the case of a fibril structure of an ECM, it acts as the template for cell behaviors: adhesion, creeping, and migration, which all lead to enhancement of new tissue formations. To mimic fibril structures, self-assembly of collagen is the main guidance in the creation of scaffolds which are similar to ECMs. Some mimicked fibril scaffolds are fabricated from self-assembly. For instance, silk fibroin, which shows performance of fibril assembly, is selected as a base material in the fabrication of some scaffolds [51, 65]. For this fabrication, silk fibroin solution is adjusted the condition: pH, temperature, and ionic strength which promotes the fibril assembly before being fabricated into scaffolds [51, 66, 67]. Some oligopeptides, which have the performance to assembly into fibril, are synthesized. These oligopeptides are prepared into a solution which is adjusted into the condition for fibril assembly [48]. In some scaffolds, they are mimicked non-homogeneous pores, which are coated with fibrils. These scaffolds show the potential to promote cell behaviors for new tissue formation [58].

From the view of surgery, mimicked morphological scaffolds need performances which fit to the operation. In the case of mimicking of morphology, which fits in with area of tissue defect, some scaffolds have to be blended with some polymers [67]. These blended scaffolds show potential to form mimicked non-homogeneous pore of ECMs [11]. Furthermore, these scaffolds have the required mechanical properties of: flexibility and strength for handling and ability for molding [68]. Some blended scaffolds show the main pores with the fibrils. These scaffolds also have mechanical strength and flexibility fitting for maxillofacial surgery [11].

### *8.2.3 Mimicked Geographically Structural 3D Scaffolds*

To mimic the geography for tissue engineering, scaffolds are mainly created with the based macro-scale of ECMs. For the damaged area, in some cases, it has the defect with multiphase tissue, which contains different macro-structures of ECMs. For instance, in some cases, mandible damaged areas of the defect include two phases of soft and bone ECMs. This defect has a thin layer of soft tissue ECM, connected to three dimensions of bone tissue ECMs. For scaffolds, that mimic geography or macro-structures, similar to ECMs, have been proposed for new tissue formation with a regular, contour shape [4, 69]. These mimicked scaffolds need the performance to maintain the space at the damage area, without collapse during new tissue formation. To mimic these scaffolds, they are fabricated with multi-steps of the process [70]. For instance, to mimic the thin layer of soft tissue ECM, scaffolds are fabricated in two dimensions, with solution casting or electro-spinning. On the other hand, to mimic the three dimensions of bone tissue ECMs, scaffolds are fabricated with particle leaching, freeze drying, or 3D printing. Finally, the 2D and 3D scaffolds are constructed to mimic the geography.

From the view of maxillofacial surgery, to mimic geography, these scaffolds have to show the basic requirements, as per the previous explanation in Sects. 8.3.1 and 8.3.2. Further performance for these scaffolds is their stable fixation at the damaged area. Construction of the size and shape for matching to the damage area is further requirement of mimicked geographical scaffolds. 3D printing is used for advanced fabrication and is often used to construct the size and shape of scaffolds matching the damaged area at the mandible [71]. To achieve these constructed scaffolds, individual 2D images of patients from X-ray instruments are used to construct 3D images by software. These individual 3D images are linked to the 3D printer, to construct 3D scaffolds which have the size and shape matching the damaged area of the patients [72].

### *8.2.4 Mimicked Physical and Mechanical 3D Scaffolds*

For maxillofacial surgery, mimicking the physical and mechanical function of ECMs is used to create scaffolds that have the performance to promote new tissue formation at the damaged area. For physical function, the swelling behavior or water uptake related to the soluble supplement of ECMs is the clues which regulate cells into different phenotypes of tissue [73]. For instance, soft tissue shows higher swelling of ECMs than bone tissue. Furthermore, the swelling behavior has the role to promote nutrient diffusion during tissue formation [73]. This is used in the guidance of creating mimicked scaffolds, so as they have physical function similar to ECMs [74]. To mimic the swelling behavior of ECM, hydrophilic polymers are often selected as the base materials to fabricate scaffolds [75].

Mechanical properties of ECMs are the clues which have an effect on the cell regulation to different phenotype of tissue [76]. An example of this is the ECM of bone tissue, which has more mechanical strength than that of soft tissue [77, 78]. High mechanical strength of the ECM is the clue for cell regulation into the phenotype of bone. Especially, the high mechanical strength has an effect on regulation of stem cells, for differentiation into the osteoblast, which is the main function for bone formation [78]. On the other hand, soft tissue has higher elasticity of the ECM than bone tissue. The higher elastic of these ECMs reveals the clue for cell regulation into phenotypes of soft tissue, in that it obtains high, extensive force [79]. To mimic mechanical strength, scaffolds are fabricated to match to the mechanical properties of ECMs [79]. For instance, scaffolds, based on high crystalline polymers which have high mechanical strength and stability, are selected to mimic bone tissue ECMs [80]. Conversely, high flexible long-chain polymers, which have high elasticity, are chosen as base materials to mimic soft tissue ECMs [81].

From the view of maxillofacial surgery, these mimicked scaffolds have to show bio-mechanic performance normally under bending force [82]. The mimicked scaffolds for the maxillofacial areas require resistant for bending force loading, without collapse after surgery [83]. This leads to maintaining of the contour shape at the damaged area without malformation. Furthermore, these scaffolds have to be created from lightweight materials, so as to avoid the collapse of the contour shape and bone resorption from the bio-mechanic performance. Mimicking of scaffolds which is similar to bio-mechanic performance at the target area is an important point in creating scaffolds for maxillofacial surgery.

### *8.2.5 Mimicked Biochemical and Biological 3D Scaffolds*

To mimic the biochemical and biological function of ECMs as well as biological signals of growth factors and cytokines called: "microenvironment", and is used to create performance scaffolds. Mimicked scaffolds, which have the biochemical and biological function similar to this microenvironment, have the potential to induce new tissue formation.

For mimicking, scaffolds are often modified with a microenvironment similar to native tissue [84]. For instance, a microenvironment is reconstituted with mixing its components under the adjusted conditions similar to native tissue [84]. Examples of components often used for this reconstitution are as follows: collagen and fibronectin [84]. The collagen and fibronectin self-assembly into fibril structure has the biochemical and biological function similar to a natural microenvironment. Self-assembled fibrils are used to fabricate scaffolds that have the function of cell recognition and biodegradability [85]. This is an important function to enhance the potential of scaffolds for activation of new tissue formation. Some examples of reconstitution of a microenvironment are the mixing of collagen, fibronectin, and growth factors to create mimicked scaffolds.

Some scaffolds are modified in conjugation with bioactive domains of microenvironments [86]. For example, RGD, YIGSR, which are the bioactive domains for cell recognition in collagen peptide and growth factor molecules, are conjugated with base synthetic or natural polymers to create mimicked scaffolds [87, 88]. These scaffolds have function to induce cell behaviors for inducing new tissue formation.

Additionally, in some cases, scaffolds are fabricated before being modified with reconstitution of a microenvironment from decellularized tissue [84]. For instance, decellularized pulp tissue from the tooth is used as biological signals for coating on silk fibroin scaffolds [84]. These coated scaffolds show the performance to induce new bone formation. These scaffolds are also proposed for surgery of bone resorption, which require biological signals to active new bone formation [89]. Nevertheless, the structure and biological function of decellularized tissue is often damaged during preparation. To solve this problem, some extracellular components of collagen and fibronectin are mixed with decellularized tissue, to reconstitution with fibril structure from collagen and binding domain from fibronectin [90]. This produces a recovered microenvironment that has both structure and function as in natural tissue.

Furthermore, the scaffolds can be further modified with a microenvironment to mimic tissue formation. For instance, the dual releasing of bone morphogenetic protein and vascular endothelial growth factors from scaffolds, based on the ECM components of: collagen or glycosaminoglycan, is created to mimic bone tissue formation [91]. This dual releasing regulates bone with blood vessel formation similar to natural tissue.

From a surgical view, these mimicked scaffolds need high biochemical and biological function to induce new tissue formation, particularly, in the case of bone resorption from diseases in elderly patients. For this case, some scaffolds have an added microenvironment with bioactive drugs for hyper-inducing of bone formation [92].

## References


1. Ceccarelli, G., Presta, R., Benedetti, L., De Angelis, M.G.C., Lupi, S.M., Baena, R.R.Y.: Emerging perspectives in scaffold for tissue engineering in oral surgery. Stem Cells Int. **2017**, 4585401 (2017)
2. Thai, T.H., Nuntanaranont, T., Kamolmatyakul, S., Meesane, J.: In vivo evaluation of modified silk fibroin scaffolds with a mimicked microenvironment of fibronectin/decellularized pulp tissue for maxillofacial surgery. Biomed. Mater. **13**, 015009 (2018)
3. Panzavolta, S., Torricelli, P., Amadori, S., Parrilli, A., Rubini, K., Bella, E., Fini, M., Bigi, A.: 3D interconnected porous biomimetic scaffolds: In vitro cell response. J. Biomed. Mater. Res. A **101**, 3560–3570 (2013)
4. Sangkert, S., Kamolmatyakul, S., Meesane, J.: Mimicked scaffolds based on coated silk woven fabric with gelatin and chitosan for soft tissue defect in oral maxillofacial area. Int. J. Artif. Organs **43**, 189–202 (2020)
5. Reddy, R., Reddy, N.: Biomimetic approaches for tissue engineering. J. Biomater Sci. Polym. Ed. **29**(14), 1667–1685 (2018)
6. Li, Y., Liu, Y., Xun, X., Zhang, W., Xu, Y., Gu, D.: Three-dimensional porous scaffolds with biomimetic microarchitecture and bioactivity for cartilage tissue engineering. ACS Appl. Mater. Interfaces **11**(40), 36359–36370 (2019)
7. Guo, J.L., Kim, Y.S., Mikos, A.G.: Biomacromolecules for tissue engineering: emerging biomimetic strategies. Biomacromol. **20**(8), 2904–2912 (2019)
8. Korpayev, S., Kaygusuz, G., Şen, M., Orhan, K., Oto, C., Karakeçili, A.: Chitosan/collagen based biomimetic osteochondral tissue constructs: A growth factor-free approach. Inter. J. Biol. Macromol. **156**, 681–690 (2020)
9. Prince, E., Kumacheva, E.: Design and applications of man-made biomimetic fibrillar hydrogels. Nat. Rev. Mater. **4**, 99–115 (2019)
10. Xiang, T., Hou, J., Xie, H., Liu, X., Gong, T., Zhou, S.: Biomimetic micro/nano structures for biomedical applications. NanoToday **2020**(35), 100980 (2020)
11. Sangkert, S., Kamolmatyakul, S., Gelinsky, M., Meesane, J.: 3D printed scaffolds of alginate/polyvinylalcohol with silk fibroin based on mimicked extracellular matrix for bone tissue engineering in maxillofacial surgery. Mater. Commun. **26**, 102140 (2021)
12. Alagoz, A.S., Hasirci, V.: 3D printing of polymeric tissue engineering scaffolds using open-source fused deposition modeling. Emergent Mater. **3**, 429–439 (2020)
13. Schipani, R., Nolan, D.R., Lally, C., Kelly, D.J.: Integrating finite element modelling and 3D printing to engineer biomimetic polymeric scaffolds for tissue engineering. Connect Tissue Res. **61**, 174–189 (2020)
14. Rodella, L.F., Favero, G., Labanca, M.: Biomaterials in maxillofacial surgery: membranes and grafts. Int. J. Biomed. Sci. **7**(2), 81–88 (2011)
15. Misch, C.M.: Maxillary autogenous bone grafting. Oral Maxillofac Surg. Clin. North Am. **23**(2), 229–238 (2011)

16. AndreasKolk, A., Handschel, J., Drescher, W., Rothamel, D., Klosse, F., Blessmann, M., Heilan, M., Wolff, K.D., Smeets, R.: Current trends and future perspectives of bone substitute materials—from space holders to innovative biomaterials. J. Cranio Maxill. Surg. **40**, 706–718 (2012)
17. Stratton-Powell, A.A., Pasko, K.M., Brockett, C.L., Tipper, J.L.: The biologic response to polyetheretherketone (PEEK) wear particles in total joint replacement: a systematic review. Clin. Orthopaed. Relat. Res. **474**, 2394–2404 (2016)
18. Davies, J.E., Matta, R., Mendes, V.C., Perri de Carvalho, P.S.: Development, characterization and clinical use of a biodegradable composite scaffold for bone engineering in oro-maxillo-facial surgery. Organogenesis **6**, 161–166 (2010)
19. Garot, C., Bettega, G., Picart, C.: Additive manufacturing of material scaffolds for bone regeneration: toward application in the clinics. Adv. Funct. Mater. **31**, 2006967 (2021)
20. Prasadh, S., Wong, R.C.W.: Unraveling the mechanical strength of biomaterials used as a bone scaffold in oral and maxillofacial defects. Oral Sci. Int. **15**, 48–55 (2018)
21. Ku, J.K., Kim, Y.K., Yun, P.Y.: Influence of biodegradable polymer membrane on new bone formation and biodegradation of biphasic bone substitutes: an animal mandibular defect model study. Maxillofac Plast Reconstr. Surg. **42**, 34 (2020)
22. Oliveira, R.L.M.S., Barbosa, L., Hurtado, C.R., Ramos, L., Oliveira, M.T.L.A., LD, Tada DB, Trichês E.: Bioglass-based scaffolds coated with silver nanoparticles: Synthesis, processing and antimicrobial activity. J. Biomed. Mater. Res. A **108**, 2447–2459 (2020)
23. Li, Y., Li, Q., Li, H., Xu, X., Fu, X., Pan, J., Wang, H., Fuh, J.Y.H., Bai, Y., Wei, S.: An effective dual-factor modified 3D-printed PCL scaffold for bone defect repair. J. Biomed. Mater. Res. B **108**, 2167–2179 (2020)
24. Dong, S., Chen, Y., Yu, L., Lin, K., Wang, X.: Magnetic hyperthermia-synergistic $H_2O_2$ self-sufficient catalytic suppression of osteosarcoma with enhanced bone-regeneration bioactivity by 3D-printing composite scaffolds. Adv. Funct. Mater. **30**, 1907071 (2020)
25. Tan, B., Tang, Q., Zhong, Y., Wei, Y., He, L., Wu, Y., Wu, J., Liao, J.: Biomaterial-based strategies for maxillofacial tumour therapy and bone defect regeneration. Int. J. Oral Sci. **13**, 9 (2021)
26. Chen, H., Yang, H., Weir, M.D., Schneider, A., Ren, K., Homayounfar, N., Oates, T.W., Zhang, K., Liu, J., Hu, T., Xu, H.H.K.: An antibacterial and injectable calcium phosphate scaffold delivering human periodontal ligament stem cells for bone tissue engineering. RSC Adv. **10**, 40157–40170 (2020)
27. He, Y., Jin, Y., Ying, X., Wu, Q., Yao, S., Li, Y., Liu, H., Ma, G., Wang, X.: Development of an antimicrobial peptide-loaded mineralized collagen bone scaffold for infective bone defect repair. Regen. Biomater. **7**, 515–525 (2020)
28. Sakkas, A., Wilde, F., Heufelder, M., Winter, K., Schramm, A.: Autogenous bone grafts in oral implantology—is it still a "gold standard"? A consecutive review of 279 patients with 456 clinical procedures. Int. J. Implant Dent. **3**, 23 (2017)
29. Zamani, Y., Amoabediny, G., Mohammadi, J., Seddiqi, H., Helder, M.N., Andieh-Doulabi, B., Klein-Nulend, J., Koolstra, J.H.: 3D-printed poly(Ɛ-caprolactone) scaffold with gradient mechanical properties according to force distribution in the mandible for mandibular bone tissue engineering. J. Mech. Behav. Biomed. **104**, 103638 (2020)
30. Zhang, W., Shi, W., Wu, S., Kuss, M., Jiang, X., Untrauer, J.B., Reid, S., Duan, B.: 3D printed composite scaffolds with dual small molecule delivery for mandibular bone regeneration. Biofabrication **12**, 035020 (2020)
31. Bartnikowski, M., Vaquette, C., Ivanovski, S.: Workflow for highly porous resorbable custom 3D printed scaffolds using medical grade polymer for large volume alveolar bone regeneration. Clin. Oral. Implant. Res. **31**, 431–441 (2020)
32. Vyas, T., Gupta, P., Kumar, S., Gupta, R., Gupta, T., Singh, H.P.: Cleft of lip and palate: A review. J. Family Med. Prim. Care **9**(6), 2621–2625 (2020)
33. Jaklová, L., Borský, J., Jurovčík, M., Hoffmannová, E., Černý, M., Dupeja, J., Velemínská, J.: Three-dimensional development of the palate in bilateral orofacial cleft newborns 1 year after early neonatal cheiloplasty: Classic and geometric morphometric evaluation. J. Cranio Maxill. Surg. **48**, 383–390 (2020)

34. Rafael, D., Luis, Z.G., Cesar Augusto, R.A., Celso Luiz, B., Cassio Eduardo, R.A.: Outcomes of surgical management of palatal fistulae in patients with repaired cleft palate. J. Craniofac Surg. **31**, e45–e50 (2020)
35. Zhang, Q., Wu, W., Qian, C., Xiao, W., Zhu, H., Guo, J., Meng, Z., Zhu, J., Ge, Z., Cui, W.: Advanced biomaterials for repairing and reconstruction of mandibular defects. Mat. Sci. Eng. C-Mater. **103**, 109858 (2019)
36. Scalzone, A., Flores-Mir, C., Carozza, D., d'Apuzzo, F., Grassia, V., Perillo, L.: Secondary alveolar bone grafting using autologous versus alloplastic material in the treatment of cleft lip and palate patients: systematic review and meta-analysis. Prog. Orthod. **6**, 20 (2019)
37. Janssen, N..G, Schreurs, R., de Ruiter, A.P., Sylvester-Jensen, H.C., Blindheim, G., Meijer, G.J., Koole, R., Vindenes, H.: Microstructured beta-tricalcium phosphate for alveolar cleft repair: a two-centre study. Int. J. Oral. Max. Surg. **48**, 708–711 (2019)
38. Ahlfeld, T., Lode, A., Richter, R.F., Pradel, W., Franke, A., Rauner, M., Stadlinger, B., Lauer, G., Gelinsky, M., Korn, P.: Toward biofabrication of resorbable implants consisting of a calcium phosphate cement and fibrin—a characterization in vitro and in vivo. Int. J. Mol. Sci. **22**(3), 1218 (2021)
39. Zheng, M., Weng, M., Zhang, X., Li1, R., Tong, Q., Chen, Z.: Beta-tricalcium phosphate promotes osteogenic differentiation of bone marrow-derived mesenchymal stem cells through macrophages. Biomed. Mater. **16**, 025005 (2021)
40. Martín-del-Campo, M., Rosales-Ibañez, R., Rojo, L.: Biomaterials for cleft lip and palate regeneration. Int. J. Mol. Sci. **20**(9), 2176 (2019)
41. Korn, P., Ahlfeld, T., Lahmeyer, F., Kilian, D., Sembdner, P., Stelzer, R., Pradel, W., Franke, A., Rauner, M., Range, U., Stadlinger, B., Lode, A., Lauer, G., Gelinsky, M.: 3D printing of bone grafts for cleft alveolar osteoplasty—in vivo evaluation in a preclinical model. Front. Bioeng. Biotechnol. **8**, 217 (2020)
42. Rees, L., Morris, P., Hall, P.: Osmotic tissue expanders in cleft lip and palate surgery: a cautionary tale. J. Plast. Reconstr. Aesthet. Surg. **61**(1), 119–120 (2008)
43. Naudot, M., Davrou, J., Djebara, A., Barre, A., Lavagen, N., Lardière, S., Azdad, S.Z., Zabijak, L., Lack, S., Devauchelle, B., Marolleau, J., Le Ricousse, S.: Functional validation of a new alginate-based hydrogel scaffold combined with mesenchymal stem cells in a rat hard palate cleft model. Plast. Reconstr. Surg. Glob. Open **8**(4), e2743 (2020)
44. Griffanti, G., Nazhat, S.N.: Dense fibrillar collagen-based hydrogels as functional osteoid-mimicking scaffolds. Int. Mater. Rev. **65**, 502–521 (2020)
45. Brito, A., Abul-Haija, Y.M., Soares da Costa, D., Novoa-Carballal, R., Reis, R.L., Ulijn, R.V., Pires, R.A., Pashkuleva, I.: Minimalistic supramolecular proteoglycan mimics by co-assembly of aromatic peptide and carbohydrate amphiphiles. Chem. Sci. **10**, 2385–2390
46. Kubyshkin, V.: Stabilization of the triple helix in collagen mimicking peptides. Org. Biomol. Chem. **17**, 8031–8047 (2019)
47. Liang, P., Zheng, J., Zhang, Z., Hou, Y., Wang, J., Zhang, C.: Bioactive 3D scaffolds self-assembled from phosphorylated mimicking peptide amphiphiles to enhance osteogenesis. J. Biomat. Sci-Polym. E **30**, 34–48 (2019)
48. Chen, J., Zou, X.: Self-assemble peptide biomaterials and their biomedical applications. Bioact. Mater. **4**, 120–131 (2019)
49. Li, J., Xing, R., Bai, S., Yan, X.: Recent advances of self-assembling peptide-based hydrogels for biomedical applications. Soft Matter. **15**, 1704–1715 (2019)
50. Wenk, E., Murphy, A.R., Kaplan, D.L., Meinel, L., Merkle, H.P., Uebersax, L.: The use of sulfonated silk fibroin derivatives to control binding, delivery and potency of FGF-2 in tissue regeneration. Biomat. **31**, 1403–1413 (2010)
51. Lu, Q., Wang, X., Lu, S., Li, M., Kaplan, D.L., Zhu, H.: Nanofibrous architecture of silk fibroin scaffolds prepared with a mild self-assembly process. Biomat. **32**, 1059–1067 (2011)
52. Bektas, C.K., Hasirci, V.: Mimicking corneal stroma using keratocyte-loaded photopolymerizable methacrylated gelatin hydrogels. J. Tissue Eng. Regen. **12**, e1899–e1910 (2018)
53. Fuchs, A., Youssef, A., Seher, A., Hochleitner, G., Dalton, P.D., Hartmann, S., Brands, R.C., Müller-Richter, U.D.A., Linz, C.: Medical-grade polycaprolactone scaffolds made by melt

electrospinning writing for oral bone regeneration—a pilot study in vitro. BMC Oral Health **19**, 28 (2019)
54. Ma, F., Xia, X., Tang, B.: Strontium chondroitin sulfate/silk fibroin blend membrane containing microporous structure modulates macrophage responses for guided bone regeneration. Carbohyd. Polym. **213**, 266–275 (2019)
55. Echave, M.C., Pimenta-Lopes, C., Pedraz, J.L., Mehrali, M., Dolatshahi-Pirouz, A., Ventur, F., Orive, G.: Enzymatic crosslinked gelatin 3D scaffolds for bone tissue engineering. Int. J. Pharmaceut **562**, 151–161 (2019)
56. Johnson, C.T., García, A.J.: Scaffold-based anti-infection strategies in bone repair. Ann. Biomed. Eng. **43**(3), 515–528 (2015)
57. Alazzawi, M., Kadim, N., Alsahib, A., Sasmazel, H.T.: Core/shell glycine-polyvinyl alcohol/polycaprolactone nanofibrous membrane intended for guided bone regeneration: development and characterization. Coatings **11**(9), 1130 (2021)
58. Jitphuthi, P., Tangtrakulwanich, B., Meesane, J.: Hierarchical porous formation, collagen and mineralized collagen modification of polylactic acid to design mimicked scaffolds for maxillofacial bone surgery. Mater. Today Commun. **13**, 46–52 (2017)
59. Deliormanlı, A.M., Atmaca, H.: Effect of pore architecture on the mesenchymal stem cell responses to graphene/polycaprolactone scaffolds prepared by solvent casting and robocasting. J. Porous Mater. **27**, 49–61 (2020)
60. Jenkins, T.L., Little, D.: Synthetic scaffolds for musculoskeletal tissue engineering: cellular responses to fiber parameters. NPI Regen. Med. **4**, 15 (2019)
61. Wang, X., Lin, M., Kang, Y.: Engineering porous β-tricalcium phosphate (β-TCP) scaffolds with multiple channels to promote cell migration, proliferation, and angiogenesis. ACS Appl. Mater. Interfaces **11**(9), 9223–9232 (2019)
62. Zhang, K., Wang, Y., Jiang, J., Wang, X., Hou, J., Sun, S., Li, Q.: Fabrication of highly interconnected porous poly(ε-caprolactone) scaffolds with supercritical $CO_2$ foaming and polymer leaching. J. Mater. Sci. **54**, 5112–5126 (2019)
63. Loh, Q.L., Choong, C.: Three-dimensional scaffolds for tissue engineering applications: role of porosity and pore size. Tissue Eng. Pt. B-Rev. **19**, 485–502 (2013)
64. Wangkulangkul, P., Jaipaew, J., Puttawibul, P., Meesane, J.: Constructed silk fibroin scaffolds to mimic adipose tissue as engineered implantation materials in post-subcutaneous tumor removal. Mater. Design **106**, 428–435 (2016)
65. Yao, D., Liu, H., Fan, Y.: Fabrication of water-stable silk fibroin scaffolds through self-assembly of proteins. RSC Adv. **6**, 61402–61409 (2016)
66. Matsumoto, A., Chen, J., Collette, A.L., Kim, U.J., Altman, G.H., Cebe, P., Kaplan, D.L.: Mechanisms of silk fibroin Sol−Gel transitions. J. Phys. Chem. B **110**(43), 21630–21638 (2006)
67. Yin, H.M., Li, X., Xu, J.Z., Zhao, B., Li, J.H., Li, Z.M.: Highly aligned and interconnected porous poly(ε-caprolactone) scaffolds derived from co-continuous polymer blends. Mater. Design **128**, 112–118 (2017)
68. Oliveira, N.K., Salles, T.H.C., Pedroni, A.C., Miguita, L., D'Ávila, M.A., Marques, M.M., Deboni, M.C.Z.: Osteogenic potential of human dental pulp stem cells cultured onto poly-ε-caprolactone/poly (rotaxane) scaffolds. Dent. Mater. **35**, 1740–1749 (2019)
69. Watcharajittanont, N., Putson, C., Pripatnanont, P., Meesane, J.: Layer-by-layer electrospun membranes of polyurethane/silk fibroin based on mimicking of oral soft tissue for guided bone regeneration. Biomed. Mater. **14**, 055011 (2019)
70. Liu, J., Zou, Q., Wang, C., Lin, M., Li, Y., Zhang, R., Li, Y.: Electrospinning and 3D printed hybrid bi-layer scaffold for guided bone regeneration. Mater. Design **210**, 110047 (2021)
71. Ciocca, L., De Crescenzio, F., Fantini, M., Scotti, R.: CAD/CAM and rapid prototyped scaffold construction for bone regenerative medicine and surgical transfer of virtual planning: A pilot study. Comput. Med. Imag. Grap. **33**, 58–62 (2009)
72. Shao, H., Sun, H., Zhang, F., Liu, A., He, Y., Fu, F., Yang, X., Wang, H., Gou, Z.: Custom repair of mandibular bone defects with 3D printed bioceramic scaffolds. J. Dent. Res. **97**(1), 68–76 (2018)

73. Chatterjee, K., Lin-Gibson, S., Wallace, W.E., Parekh, S.H., Lee, Y.J., Marcus, T., Cicerone, M.T., Young, F.Y., Simon, C.G., Jr.: The effect of 3D hydrogel scaffold modulus on osteoblast differentiation and mineralization revealed by combinatorial screening. Biomat. **31**, 5051–5062 (2010)
74. Thangprasert, A., Tansakul, C., Thuaksubun, N., Meesane, J.: Mimicked hybrid hydrogel based on gelatin/PVA for tissue engineering in subchondral bone interface for osteoarthritis surgery. Mater. Design **183**, 108113 (2019)
75. Riva, R., Shah, U., Thomassin, J.M., Yilmaz, Z., Lecat, A., Colige, A., Jérôme, C.: Design of degradable polyphosphoester networks with tailor-made stiffness and hydrophilicity as scaffolds for tissue engineering. Biomacromol. **21**(2), 349–355 (2020)
76. Evans, N.D., Gentleman, E.: The role of material structure and mechanical properties in cell–matrix interactions. J. Mater. Chem. B **2**, 2345–2356 (2014)
77. Padhi, A., Nain, A.S.: ECM in differentiation: a review of matrix structure, composition and mechanical properties. Ann. Biomed. Eng. **48**, 1071–1089 (2020)
78. Smith, L.R., Cho, S., Discher, D.E.: Stem cell differentiation is regulated by extracellular matrix mechanics. Physiology (Bethesda) **33**(1), 16–25 (2018)
79. Chaudhuri, O., Cooper-White, J., Janmey, P.A., Mooney, D.J., Shenoy, V.B.: Effects of extracellular matrix viscoelasticity on cellular behavior. Nature **584**, 535–546 (2020)
80. Dwivedi, R., Kumar, S., Pandey, R., Mahajan, A., Nandana, D., Katti, D.S., Mehrotra, D.: Polycaprolactone as biomaterial for bone scaffolds: Review of literature. J. Oral Biol. Craniofac. Res. **10**(1), 381–388 (2020)
81. Parrag, I.C., Woodhouse, K.A.: Development of biodegradable polyurethane scaffolds using amino acid and dipeptide-based chain extenders for soft tissue engineering. J. Biomat. Sci-Polym. E **21**, 843–862 (2010)
82. Cheng, K., Liu, Y., Wang, J.H., Jun, J.C., Jiang, X., Wang, R., Baur, D.A.: Biomechanical behavior of mandibles reconstructed with fibular grafts at different vertical positions using finite element method. J. Plast. Reconstr. Aes. **72**, 281–289 (2019)
83. Hu, J., Wang, J.H., Wang, R., Yu, X.B., Liu, Y., Baur, D.A.: Analysis of biomechanical behavior of 3D printed mandibular graft with porous scaffold structure designed by topological optimization. 3D Print Med. **5**, 5 (2019)
84. Sangkert, S., Kamonmattayakul, S., Chai, W.L., Meesane, J.: Modified porous scaffolds of silk fibroin with mimicked microenvironment based on decellularized pulp/fibronectin for designed performance biomaterials in maxillofacial bone defect. J. Biomed. Mater. Res. A **105**, 1624–1636 (2017)
85. Ilamaran, M., Janeena, A., Valappil, S., Ramudu, K.N., Shanmugam, G., Niraikulam, A.A.: self-assembly and higher order structure forming triple helical protein as a novel biomaterial for cell proliferation. Biomater. Sci. **7**, 2191–2199 (2019)
86. Mohamed, M.A., Fallahi, A., El-Sokkary, A.M.A., Salehi, S., Akl, M.A., Jafari, A., Tamayol, A., Fennirih, H., Khademhosseini, A., Andreadis, S.T., Cheng, C.: Stimuli-responsive hydrogels for manipulation of cell microenvironment: From chemistry to biofabrication technology. Prog. Polym. Sci. **98**, 101147 (2019)
87. Philip, D.L., Silantyeva, E.A., Becker, M.L., Willits, R.K.: RGD-Functionalized Nanofibers Increase Early GFAP Expression during Neural Differentiation of Mouse Embryonic Stem Cells. Biomacromol **20**(3), 1443–1454 (2019)
88. Su, J., Satchell, S.C., Wertheim, J.A., Shah, R.N.: Poly(ethylene glycol)-crosslinked gelatin hydrogel substrates with conjugated bioactive peptides influence endothelial cell behavior. Biomaterials **201**, 99–112 (2019)
89. Dai, J., Qiao, W., Shi, J., Liu, C., Hu, X., Dong, N.: Modifying decellularized aortic valve scaffolds with stromal cell-derived factor-1$\alpha$ loaded proteolytically degradable hydrogel for recellularization and remodeling. Acta. Biomater. **88**, 280–292 (2019)
90. Sangkert, S., Kamonmattayakul, S., Chai, W.L.: Meesane J*, A biofunctional-modified silk fibroin scaffold with mimic reconstructed extracellular matrix of decellularized pulp/collagen/fibronectin for bone tissue engineering in alveolar bone resorption. Mater. Lett. **166**, 30–34 (2016)

91. Dou, D.D., Zhou, G., Liu, H.W., Zhan, J., Liu, M.L., Xiao, X.F., Fei, J.J., Guan, X.L., Fan, Y.B.: Sequential releasing of VEGF and BMP-2 in hydroxyapatite collagen scaffolds for bone tissue engineering: Design and characterization. Int. J. Biol. Macromol. **123**, 622–628 (2019)
92. He, F., Lu, T., Fang, X., Tian, Y., Li, Y., Zuo, F., Ye, J.: Modification of honeycomb bioceramic scaffolds for bone regeneration under the condition of excessive bone resorption. J. Biomed. Mater. Res. Part A **107A**, 1314–1323 (2019)

# Chapter 9
# Mimicked 2D Scaffolds for Maxillofacial Surgery

In maxillofacial surgery, there are many cases which have defects in bone including soft tissue [1]. These defects normally come from either trauma or diseases [2]. In the case of soft tissue defects, because of physical and biological function, there is a requirement for different scaffolds from bone tissue. For instance, soft tissue shows physical function of higher elasticity and elongation than bone tissue [2]. Furthermore, soft tissue has a faster repairing time than bone tissue [3–5]. Therefore, biomaterials which match these functions are required for soft tissue in surgery. Mimicking is the approach to create biomaterials that have similar functions as that of soft tissue [6, 7]. Created scaffold biomaterials which have similar physical and biological function to soft tissue are normally the key for mimicking [6]. For instance, elastic polymers conjugated with biological signals are fabricated into scaffolds, based on the mimicking of physical and biological function of soft tissue [3]. Moreover, in some cases scaffolds need an antimicrobial function for surgery [8]. For example, antimicrobial drugs are added into the scaffolds during fabrication [9]. These scaffolds are places to cover over defect site before closure to avoid microbial invasion.

Maxillofacial surgery which involves a soft tissue defect is demonstrated. Mimicked scaffolds for defects included soft tissue in the maxillofacial are shown. Some examples of mimicked scaffolds in the maxillofacial are also explained. Furthermore, some ideas to develop mimicked scaffolds for soft tissue in the maxillofacial are described in this chapter.

## 9.1 2D Scaffolds for Maxillofacial Surgery

For surgery in the maxillofacial, the crucial problem is the interference of new soft tissue formations that disturbs new bone formation. This leads to incomplete, new bone formation [10]. In such cases, surgeons need to place another barrier membrane



J. Meesane, *Mimicked Tissue Engineering Scaffolds for Maxillofacial and Articular Cartilage Surgery*, Engineering Materials,
https://doi.org/10.1007/978-981-19-7830-2_9

scaffold that induces the formation of new soft tissue [3, 11]. Furthermore, the membrane has the performance to prevent the invasion of cells from the soft tissue that disturbs new bone formation. For good membrane scaffolds, the cells of the soft tissue do not penetrate into the bony part. On the other hand, poor membrane scaffolds lead to cell penetration, which is the cause of incomplete bone tissue formation. Therefore, the scaffold in contact with soft tissue needs to have a barrier function, so as to prevent cell penetration to the bone. This requires a biological function to promote soft tissue formation, while on the other side of the scaffold, it needs to promote bone tissue regeneration [12]. Furthermore, the scaffold has to show a suitable degradation timeline that fits the regeneration of both soft tissue and bone tissue.

The main requirements to design scaffolds fitting maxillofacial surgery are classified into two parts: (1) structure and (2) function. The structure of the scaffolds needs pore sizes which are suitable for cell behaviors: adhesion, proliferation, and migration [3]. Scaffolds with suitable geometry and sizes support these cell behaviors, which regulate new tissue formation, with the regular, recovered contour shape of the lip [7, 13]. In the case of non-suitable sizes, cell behaviors show on only the surface of scaffolds. This leads to an effect on irregular new tissue formation and recovered contour shape of the lip [13]. Hence, both the geometry and size of said scaffolds are required for fitting to the contour shape at the defect site [14]. These scaffolds assist to maintain the regular form of the lip after an operation. In the case of non-fit geometry and size, the scaffolds guide the new tissue formation into an irregular, recovered contour shape, which leads to the dysfunction of the lip [13].

For the function of scaffolds, they need mechanical stability, hydrophilicity, biodegradability, biocompatibility, and non-toxicity [15, 16]. First, mechanical stability is the function to maintain the contour shape at the defect site, until the completed new tissue formation has taken place [17, 18]. This leads to a regular, recovered contour shape of the lip. On the other hand, insufficient mechanical stability causes the scaffold to collapse during new tissue formation [18]. This has an effect on non-completed new tissue formation and irregular, recovered contour shape of the lip [14].

Second, hydrophilicity is the function supporting the attachment of some secreted biological signals from cells and the nutrients to promote new tissue formation [19]. Furthermore, hydrophic scaffolds act as a suitable template for inducing cell behaviors: adhesion, spreading, migration, and proliferation [20]. This leads to the enhancement of new tissue formation [19]. Hydrophilicity is an important parameter affecting the increasing of biodegradation, which is the other requirement of scaffolds [21].

Third, biodegradability is an important function to assist the promotion of regular, new tissue formation [22]. The degradation rate must match to the time period of new tissue formation [23]; otherwise, the tissue shows irregular morphology and dysfunction. In cases such as these, patients may require a second operation.

Fourth, biocompatibility is the function which supports cell behaviors: adhesion, proliferation, spreading, migration, and differentiation to enhance new tissue formation [24]. For large defects, biocompatible and bioactive materials assist with the promotion of these cell behaviors for activation of new tissue formation [25].

Fifth, non-toxicity is another important function of scaffolds, so as to avoid side effects after an operation [26]. The suitable selection of materials is the key to avoid toxicity during and after tissue engineering. The suitable materials are normally hydrophilic polymers, which degraded without any residual toxic agents [17, 26]. For some cases, these toxic agents are the cause of irregular tissue formations or diseases [27, 28]. Furthermore, avoiding base materials with many chemical additives is an important point in the selection of suitable polymers for scaffolds.

### *9.1.1 2D Scaffolds for Guided Bone Regeneration*

In some cases, the area of the teeth removed shows bone loss which have insufficient volume for dental implantation. In these cases, the patient needs surgery, consisting of increasing volume at the area of bone loss. Guided bone regeneration (GBR) is an effective surgical approach, which is often used to increase the bone volume before dental implantation (Fig. 9.1). In this approach, the bone matrix compound, for instance hydroxyapatite, tri-calcium phosphate, biphasic calcium phosphate, and other bioactive compounds, is added into the area of bone loss [29, 30]. Then, the bone matrix compound, added area, is covered with a membrane that normally has a barrier function, so as to prevent soft tissue invasion during new bone formation. Next, the membrane-covered area obtains closure with suturing. After complete new bone tissue formation, dental implantation is then undertaken.

In this GBR approach, 2D scaffolds are used as the performance barrier membranes to assist in the regulated, new bone formation [29, 31]. These scaffolds are designed into the porous membrane, which prevent soft tissue invasion during

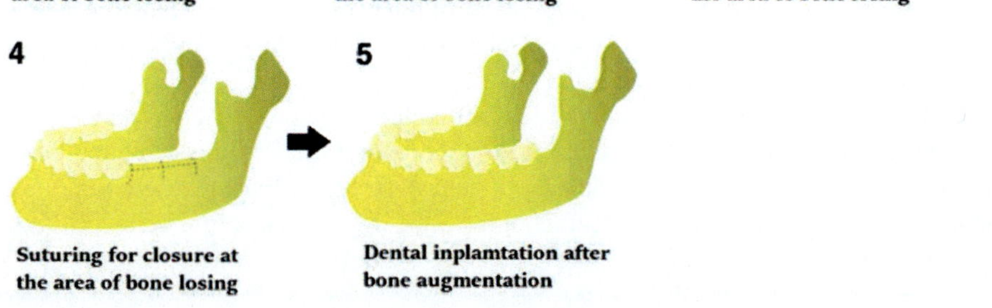


**Fig. 9.1** Guided bone regeneration of bone loss areas at the mandible

bone formation. Furthermore, they also have a permeable function for nutrient and biological molecule [32]; additionally, 2D membrane scaffolds need the structure and properties which serve complete new bone formation. These scaffolds show osseointegration. Collagen is often used as a GBR membrane, because it has good function to promote new tissue formation [33, 34]. Nevertheless, this membrane has the weak point of a fast rate of biodegradation, leading to insufficient mechanical strength to hold its contour shape during new tissue formation. Some mechanical stable scaffolds, with low biodegradation, have been developed to substitute collagen [33]. Some scaffolds have high function for inducing new bone formation and blood vessel formation [35], and for some cases, they are also developed with an antimicrobial function [36]. Importantly, to design these scaffolds, one has to strongly consider the views of their potential for new tissue formation and surgical performance.

### 9.1.2 2D Scaffolds for Alveolar Cleft Lip and Palate

For alveolar cleft surgery, includes patients who have these defects: (1) lip or (2) lip including palate. For alveolar cleft lip, the defect is concluded with multilayers of soft tissue: epithelium, lamina propria, and submucosa [37]. In surgery to correct this defect, surgeons normally use the technique of flapping [38]. However, for some cases with large defects, surgeons need biomaterial, to recover the regular shape around the oral area [39]. These biomaterials are required to have stability during tissue repairing. Furthermore, they need performance based on the design fitting to the operation, for instance ability for molding, sufficient stability to handle, tear resistant for suturing, and so forth [14].

In some cases, surgeons use 2D scaffolds as performance biomaterials with bioactivity and resorbability, to promote new tissue formation at the defect area (Fig. 9.2). 2D scaffolds mainly act as structural materials to promote cell behaviors, leading to new tissue formation. For 2d scaffolds, they further need sufficient strength to maintain the contour shape without collapse after surgery and during new tissue formation [2, 15]. They also required the advance function of inducing new soft tissue formations [40].

In case of the defect including total soft and partial bone areas, surgeons add bone matrix compounds or 3D scaffolds without flapping (Fig. 9.3). Then, the added defect of the partial bone area is covered with a 2D scaffold leaved in the soft tissue area. Afterward, the defect is sutured.

In the case of cleft lip and palate, the defect includes the total soft and bone tissue area which is operated on with flapping, for the preparation of space for the bone matrix compound, or 3D scaffold adding (Fig. 9.4). Next, the bone matrix compound or 3D scaffolds are added into the defect of the bone tissue area, before being covered with the 2D scaffolds left in the soft tissue area; the defect is then sutured.

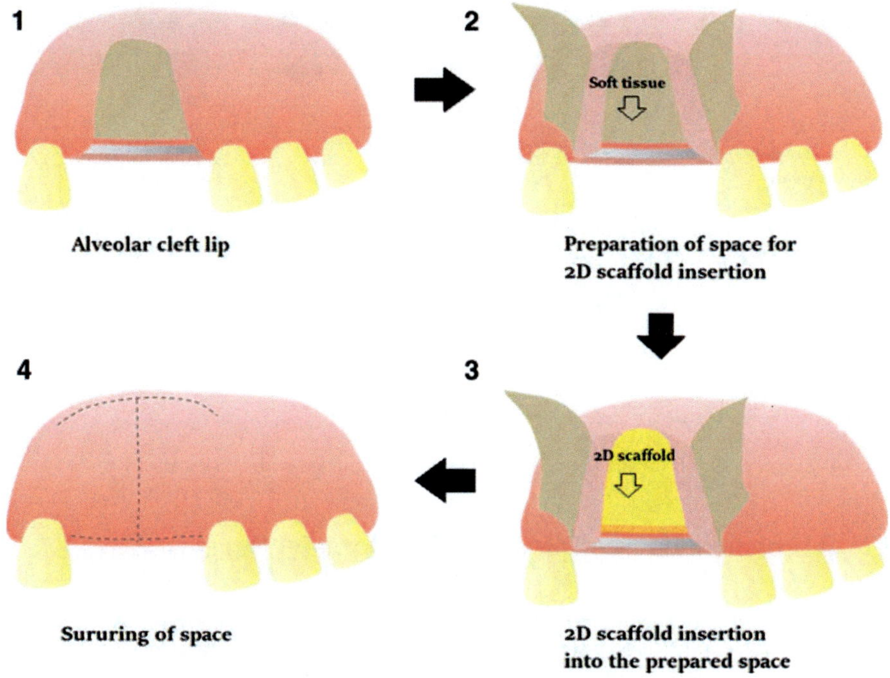


**Fig. 9.2** Surgery of alveolar clef lip with assistance of a 2D scaffold

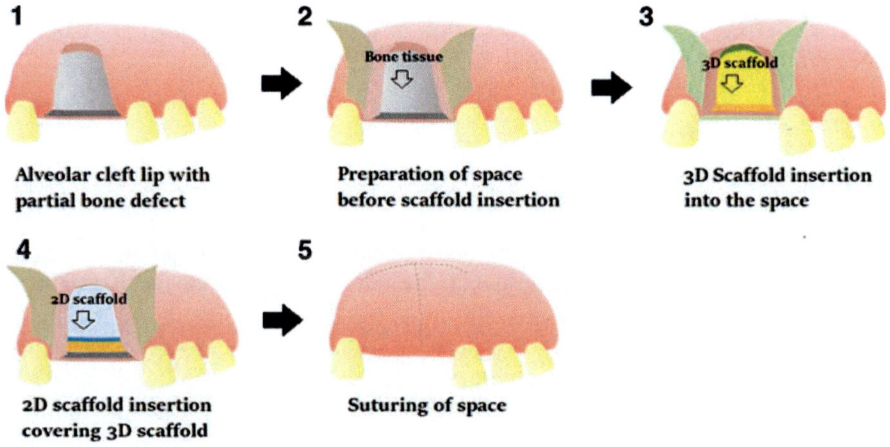


**Fig. 9.3** Surgery of cleft lip with partial bone defect by a 3D scaffold assisted with a 2D scaffold

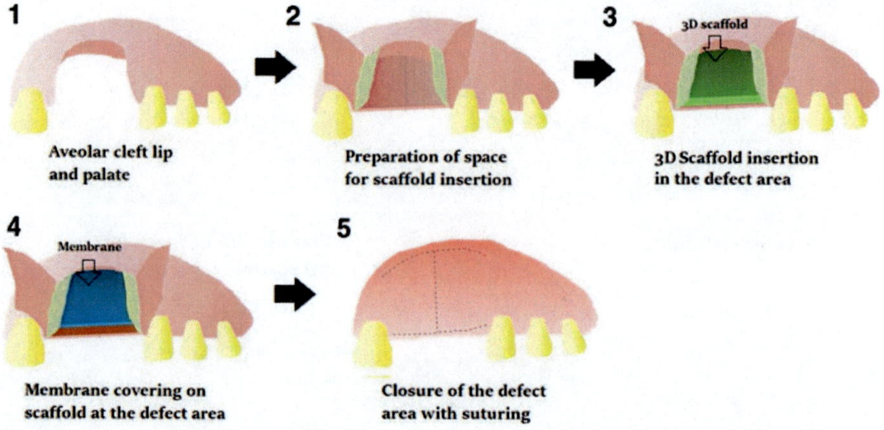


**Fig. 9.4** Surgery of cleft lip and palate, by a 3D scaffold, assisted with a 2D scaffold

## 9.2 Mimicked 2D Scaffolds for Maxillofacial Surgery

In maxillofacial surgery, mimicked 2D scaffolds are the performance biomaterials assisting in completion of new tissue formation. Based on the view of materials, mimicked 2D scaffolds are created into similar structures, with similar function of ECMs in natural soft tissue. From the view of surgery, mimicked 2D scaffolds have to perform as barrier membranes, which are in contact between soft and bone tissue. Therefore, to succeed in the creation of mimicked 2D scaffold, one needs optimization of those views: materials and surgery.

### *9.2.1 Structural Mimicking in Scaffolds*

Structural mimicking of scaffolds is divided into two groups: (1) nano-structures, (2) microstructures, and macro-structures [41]. In the development of nano-structure, scaffolds are fabricated with the concept of (1) top-down or large to small and (2) bottom-up or small to large [42]. For instance, in the case of top-down scaffolds these are fabricated with electro-spinning. In this fabrication, the polymer solution is used as the raw material, which is the top or large scale. The polymer solution is spun into nano-fibrils or small scale, under an electric field. For the bottom-up, some molecules are used to assemble into the nano-structure. For example, collagen or oligopeptide molecules are used to assemble into nano-fibrils [43]. Nano-fibrils are the structures which are then used in the construct of these mimicked scaffolds. These fibrils are mimicked similar to ECMs in tissue. The mimicked structure of ECMs, which has core fibril of collagen connected to small fibrils of the other fibrillar proteins, electro-spun fibrils are fabricated for optimized condition [44]. This leads to the formation core fibrils with split small fibrils, which are then similar to ECMs.

Another example is by collagen fibril assembly with mineralization of calcium phosphate being used to mimic the nano-structure of bone ECMs [45]. In this mimicking, collagen molecules self-organize into nano-fibrils that have regular, small spaces. These spaces are deposited with calcium phosphate crystals from mineralization [46].

Mimicking on microstructures of scaffolds is mainly focused on their porous structure that is similar to natural bone and is classified into two porous structures of (1) cortical and (2) trabecular bone. The cortical bone shows a more densely morphology with small pores than that of trabecular bone. Pore size and the structure of the bone are mimicked for scaffold fabrication. Pore sizes of the bone are in the range of micrometer [47, 48]. For porous structures, natural bone demonstrates the hierarchical structure, which has different pore sizes. First, to mimic the pore size of the bone, freeze drying and particle leaching are simply used to fabricate the scaffolds [49]. Second, in the case of a porous structure, combined techniques are used to fabricate scaffolds similar to natural bone [49]. For instance, particle leaching is used to fabricate scaffolds which have large pores, and then, the particle leached scaffolds are immersed into a polymer solution, before freeze drying is undertaken to form small pores. Finally, these scaffolds show hierarchical porous structures of both the main, large pores incorporated with small pores [50, 51]. They are also used to other tissues [50]. Furthermore, to mimic pore size, 3D printing is the performance technique for porous scaffolds, which have pore sizes similar to natural bone [51].

To mimic a macro-structure, single layer, 2D scaffolds demonstrate a simple structure and ease for fabrication [2, 3]. Solution casting and electro-spinning are generally the approach for fabrication of single layer, 2D scaffolds [2, 29, 52]. Mimicking both the structure and function, similar to soft tissue, is an important requirement [2, 29]. In the soft tissue at the alveolar cleft lip, the structure is composed with collagen type I and proteoglycan that self-organizes into the thin layer of the extracellular matrix (ECM) [53]. ECMs have structures of collagen fibrils embedded into the basement of proteoglycan. The properties of the ECM depend on the structure and function of those components. For instance, ECMs with regular formation of collagen fibrils show higher mechanical strength than irregular formation [54]. Mainly, the ECM of soft tissue at oral and maxillofacial area acts as the template for epithelial cells to grow into regular tissue [55].

To mimic the structure similar to the ECM, single layer 2D scaffolds are made from the assembled protein fibrils, embedded into the polysaccharide hydrogel [56]. For example, selected silk fibroin is prepared into assembled fibrils that are similar to collagen type I [57]. For the selected polysaccharide, alginate is an example to mimic similar function to proteoglycan in ECMs [42]. Based on this mimicking, other natural and synthetic polymers are used to substitute silk fibroin and alginate [58]. Furthermore, the constructed fibrils with performance fabrication, particularly in electro-spinning, are used to mimic the single layer, 2D scaffolds similar to the ECM [44]. From the view of operation, single layer, 2D scaffolds need to have flexibility for molding and sufficient dimensions to fit with the defect area. Moreover, they must have sufficient tear strength for suturing [3, 44].

For natural soft tissue at the maxillofacial area, its structure shows multilayers: epithelium, lamina propria, and submucosa [53]. This has suitable function to maintain the contour shape. Additionally, this multilayer connects to the layer of bone. The multilayers of alveolar lip have ECM membranes for cell attachment. To mimic the multilayers, 2D scaffolds are constructed and then used to develop the performance, similar to alveolar lip. Each layer is normally fabricated with electro-spinning, so as to construct the fibrous structure similar to this ECM [3]. Selection of the proper polymer materials is the key to mimic these multilayers, as the selected polymer materials must firstly have physical performance which is suitable to absorb nutrients during tissue formation. Secondly, the polymer materials must show mechanical performance: strength, toughness, and elasticity similar to that of natural ECMs. This is the clue to guide cell behaviors for regular tissue formation. Thirdly, they must have biological performance for cell recognition or binding similar to the natural components in the ECM. This regulates the cells to organize tissue into a regular form.

For instances of multilayer mimicked fibrous 2D scaffolds, elastic electro-spun polyurethane (PU) is constructed for the first layer and then covered with a second layer of high strength electro-spun silk fibroin (SF) [3]. After which, the bilayers are covered with electro-spun SF. This construction has the multilayers of PU/SF/PU. Based on this construction, other polymers, which have the performance similar to the ECM, can be chosen to mimic scaffolds instead.

### *9.2.2 Mimicked Function in Scaffolds*

An important function of scaffolds is osteoinduction, and there are many researches which are mainly focused on the development of osteoinduction of scaffolds fitting to maxillofacial surgery [29, 59]. Those researches incorporate bioactive molecules that have important roles to induce cell behaviors which enhance bone tissue formation. Bioactive molecules are usually growth factors incorporated into mimicked scaffolds, by either a physical or chemical approach [60–62]. Bone morphogenetic proteins (BMPs) are the trigger to induce osteoblast cells to promote tissue formation [63, 64]. Transforming growth factor-β (TGF-β) is another growth factor used to enhance bone formation [65]. Furthermore, vascular endothelial growth factor (VEFG) is the growth factor to active blood vessel formation [66]. The combination of BMP2 with VEGF is used to create mimicked scaffolds, which show dual function to activate bone and blood vessel formation [67]. This leads to mimicked bone similar to natural tissue.

Combination of mimicked scaffolds, with bioactive inorganic compounds, is used to develop osteoinduction [29, 68]. Hydroxyapatite (HA) is the main compound incorporated with scaffolds, because it has chemical groups in its molecules that fit to active cell behaviors [69]. Magnesium or its derivative molecules have been proposed for application to induce bone formation [70]. Strontium (Sr) or its derivative molecules are the bioactive molecules used to incorporate with scaffolds [71].

Mainly, these molecules show the function to induce bone formation; hence, scaffolds with strontium are applied for osteoporosis treatment [72].

Incorporation of growth factors with other bioactive molecules, which have specific functions, the treat bone defects are an attractive to develop mimicked scaffolds [73]. An example of this is antimicrobial components, which are added into the scaffolds coupled with bioactive molecules [74]. These scaffolds act as a dual function of inducing bone formation and antimicrobial performance, and so these scaffolds are proposed for application to surgery incorporated with osteomyelitis treatment [75].

In this chapter, the instances of mimicked 2D scaffolds were described for their potential in the application for maxillofacial surgery. Mimicked scaffolds were explained from material and surgical views. In concerns to the material view, structure and function that are important to mimic scaffolds, similar to natural tissue at the maxillofacial area, were emphasized. In terms of the surgical view, creating mimicked scaffolds, which are fitted into the defect sites and are easy to handle during an operation, was explained.

## References

1. Harris, C.M., Laughlin, R.: Reconstruction of hard and soft tissue maxillofacial defects. Atlas Oral. Maxillofac. Surg. Clin. **21**(1), 127–138 (2013)
2. Kumar, R.V.K., Devireddy, S.K., Gali, R.S., Chaithanyaa, N., Sridhar. A clinician's role in the management of soft tissue injuries of the face: a clinical paper. J. Maxillofac. Oral Surg. **12**(1), 21–29 (2013)
3. Sangkert, S., Kamolmatyakul, S., Meesane, J.: Mimicked scaffolds based on coated silk woven fabric with gelatin and chitosan for soft tissue defect in oral maxillofacial area. Int. J. Artif. Organs **43**(3), 189–202 (2020)
4. Sybil, D., Sawai, M., Faisal, M., Singh, S., Jain, V.: Platelet-rich fibrin for hard- and soft-tissue healing in mandibular third molar extraction socket. Ann. Maxillofac. Surg. **10**(1), 102–107 (2020)
5. Chouhan, D., Dey, N., Bhardwaj, N., Mandal, B.B.: Emerging and innovative approaches for wound healing and skin regeneration: Current status and advances. Biomaterials **216**, 119267 (2019)
6. Haffner-Luntzer, M., Hankenson, K.D., Ignatius, A., Pfeifer, R., Khader, B.A., Hildebrand, F., van Griensven, M., Pape, H.C., Lehmicke, M.: Review of animal models of comorbidities in fracture-healing research. J. Orthop. Res. **37**, 2491–2498 (2019)
7. Toledano, M., Toledano-Osorio, M., Carrasco-Carmona, A., Vallecillo, C., Lynch, C.D., Osorio, M.T., Osorio, R.: State of the art on biomaterials for soft tissue augmentation in the oral cavity. Part I: natural polymers-based biomaterials. Polym. (Basel) **12**(8), 1850 (2020)
8. Watcharajittanont, N., Putson, C., Pripatnanont, P., Meesane, J.: Layer-by-layer electrospun membranes of polyurethane/silk fibroin based on mimicking of oral soft tissue for guided bone regeneration. Biomed. Mater. **14**, 055011 (2019)
9. Sharif, F., Tabassum, S., Mustafa, W., Asif, A., Zarif, F., Tariq, M., Siddiqui, S.A., Gilani, M.A., Rehman, I.U., MacNeil, S.: Bioresorbable antibacterial PCL-PLA-nHA composite membranes for oral and maxillofacial defects. Polym. Compos. **40**, 1564–1575 (2019)
10. Lu, H., Liu, Y., Guo, J., Wu, H., Wang, J., Wu, G.: Biomaterials with antibacterial and osteoinductive properties to repair infected bone defects. Int. J. Mol. Sci. **17**(3), 334 (2016)

11. Dimitriou, R., Mataliotakis, G.I., Calori, G.M., Giannoudis, P.V.: The role of barrier membranes for guided bone regeneration and restoration of large bone defects: current experimental and clinical evidence. BMC Med. **10**, 81 (2012)
12. Berton, F., Porrelli, D., Di Lenarda, R., Turco, G.: A critical review on the production of electrospun nanofibres for guided bone regeneration in oral surgery. Nanomaterials **10**(1), 16 (2020)
13. Saigo, L., Kumar, V., Liu, Y., Lim, J., Teoh, S.H., Goh, B.T.: A pilot study: Clinical efficacy of novel polycaprolactone-tricalcium phosphate membrane for guided bone regeneration in rabbit calvarial defect model. J. Oral Maxil. Surg. Med. Pathol. **30**, 212–219 (2018)
14. Martín-del-Campo, M., Rosales-Ibañez, R., Rojo, L.: Biomaterials for Cleft Lip and Palate Regeneration. Int. J. Mol. Sci. **20**(9), 2176 (2019)
15. Korn, P., Ahlfeld, T., Lahmeyer, F., Kilian, D., Sembdner, P., Stelzer, R., Pradel, W., Franke, A., Rauner, M., Range, U., Stadlinger, B., Lode, A., Lauer, G., Gelinsky, M.: 3D printing of bone grafts for cleft alveolar osteoplasty – in vivo evaluation in a preclinical model. Front. Bioeng. Biotechnol. **8**, 217 (2020)
16. Sharif, F., Ur Rehman, I., Muhammad, N., MacNeil, S.: Dental materials for cleft palate repair. Mat. Sci. Eng. C Mater. **61**, 1018–1028 (2016)
17. Qu, H., Fu, H., Hana, Z., Sun, Y.: Biomaterials for bone tissue engineering scaffolds: a review. RSC Adv. **9**, 26252–26262 (2019)
18. Maroulakos, M., Kamperos, G., Tayebi, T., Halazonetis, D., Ren, Y.: Applications of 3D printing on craniofacial bone repair: a systematic review. J. Dent. **80**, 1–14 (2019)
19. Shen, C., Witek, L., Flores, R.L., Tovar, N., Torroni, A., Coelho, P.G., Kasper, F.K., Wong, M., Young, S.: Three-dimensional printing for craniofacial bone tissue engineering. Tissue Eng. Pt A **26**, 1303–1311 (2020)
20. Amani, H., Arzaghi, H., Bayandori, M., Dezfuli, A.S., Pazoki-Toroudi, H., Shafiee, A., Moradi, L.: Controlling cell behavior through the design of biomaterial surfaces: a focus on surface modification techniques. Adv. Mater. Interfaces **6**, 1900572 (2019)
21. Rashad, A., Mohamed-Ahmed, S., Ojansivu, M., Berstad, K., Yassin, M.A., Kivijärvi, T., Heggset, E.B., Syverud, K., Mustafa, K.: Coating 3D printed polycaprolactone scaffolds with nanocellulose promotes growth and differentiation of mesenchymal stem cells. Biomacromol. **19**(11), 4307–4319 (2018)
22. Shuai, C., Yang, W., Feng, P., Peng, S., Pan, H.: Accelerated degradation of HAP/PLLA bone scaffold by PGA blending facilitates bioactivity and osteoconductivity. Bioact. Mater. **6**, 490–502 (2021)
23. Ma, Y., Hu, N., Liu, J., Zhai, X., Wu, M., Hu, C., Li, L., Lai, Y., Pan, H., Lu, W.W., Zhang, X., Luo, Y., Ruan, C.: Three-dimensional printing of biodegradable piperazine-based polyurethane-urea scaffolds with enhanced osteogenesis for bone regeneration. ACS Appl. Mater. Interfaces **11**(9), 9415–9424 (2019)
24. Lu, H.T., Lu, T.W., Chen, C.H., Mi, F.L.: Development of genipin-crosslinked and fucoidan-adsorbed nano-hydroxyapatite/hydroxypropyl chitosan composite scaffolds for bone tissue engineering. Int. J. Biol. Macromol. **128**, 973–984 (2019)
25. Kazimierczak, P., Benko, A., Nocun, M., Przekora, A.: Novel chitosan/agarose/hydroxyapatite nanocomposite scaffold for bone tissue engineering applications: comprehensive evaluation of biocompatibility and osteoinductivity with the use of osteoblasts and mesenchymal stem cells. Int. J. Nanomed. **14**, 6615–6630 (2019)
26. Dasgupta, S., Maji, K., Nandi, S.K.: Investigating the mechanical, physiochemical and osteogenic properties in gelatin-chitosan-bioactive nanoceramic composite scaffolds for bone tissue regeneration: in vitro and in vivo. Mater. Sci. Eng. C Mater. **94**, 713–728 (2019)
27. Nezhad-Mokhtari, P., Ghorbani, M., Roshangar, L., Rad, J.S.: A review on the construction of hydrogel scaffolds by various chemically techniques for tissue engineering. Eur. Polym. J. **17**, 64–76 (2019)
28. Kim, K.T., Eo, M.Y., Nguyen, T.T.H., Kim, S.M.: General review of titanium toxicity. Int. J. Implant Dent. **5**, 10 (2019)

29. Hall, N.J., Samelko, L., Hammond, D.: the inflammatory effects of breast implant particulate shedding: comparison with orthopedic implants. Aesthet. Surg. J. **39**, S36–S48 (2019)
30. Watcharajittanont, N., Tabrizian, M., Putson, C., Pripatnanont, P., Meesane, J.: Osseointegrated membranes based on electro-spun TiO2/hydroxyapatite/polyurethane for oral maxillofacial surgery. Mater. Sci. Eng. C Mater **108**, 110479 (2020)
31. Lei, B., Guo, B., Rambhia, K.J., Ma, P.X.: Hybrid polymer biomaterials for bone tissue regeneration. Front. Med. **13**, 189–201 (2019)
32. Deraine, A., Rebelo Calejo, M.T., Agniel, R., Kellomäki, M., Pauthe, E., Boissière, M., Massera, J.: Polymer-based honeycomb films on bioactive glass: toward a biphasic material for bone tissue engineering applications. ACS Appl. Mater. Interfaces **13**(25), 29984–29995 (2021)
33. Wang, J., Wang, L., Zhou, Z., Lai, H., Xu, P., Liao, L., Wei, J.: Biodegradable polymer membranes applied in guided bone/tissue regeneration: a review. Polym. (Basel) **8**(4), 115 (2016)
34. Lee, S.W., Kim, S.G.: Membranes for the guided bone regeneration. Maxillofac. Plast. Reconstr. Surg. **36**(6), 239–246 (2014)
35. Sbricoli, L., Guazzo, R., Annunziata, M., Gobbato, L., Bressan, E., Nastri, L.: Selection of collagen membranes for bone regeneration: a literature review. Mater. (Basel) **3**(3), 786 (2020)
36. Kaigler, D., Silva, E.A., Mooney, D.J.: Guided bone regeneration using injectable vascular endothelial growth factor delivery gel. J. Periodontol. **84**, 230–238 (2013)
37. Chen, P., Wu, Z., Leung, A., Chen, X., Landao-Bassonga, E., Gao, J., Chen, L., Zheng, M., Yao, F., Yang, H., Lidgren, L., Allan, B., Liu, Y., Wang, T., Zheng, M.: Fabrication of a silver nanoparticle-coated collagen membrane with anti-bacterial and anti-inflammatory activities for guided bone regeneration. Biomed. Mater. **13**, 065014 (2018)
38. Kyung, H., Kang, N.: Management of alveolar cleft. Arch. Craniofac. Surg. **16**(2), 49–52 (2015)
39. Xu, J., Yao, J.: Repair of alveolar cleft. In: Yao, J., Xu, J. (eds.) Atlas of cleft lip and palate & facial deformity surgery, Springer, Singapore (2020)
40. Berger, M., Probst, F., Schwartz, C., Cornelsen, M., Seitz, H., Ehrenfeld, M., Otto, S.: A concept for scaffold-based tissue engineering in alveolar cleft osteoplasty. J. Cranio Maxill. Surg. **43**, 830–836 (2015)
41. Rodríguez-Méndez, I., Fernández-Gutiérrez, M., Rodríguez-Navarrete, A., Rosales-Ibáñez, R., Benito-Garzón, L., Vázquez-Lasa, B., Román, J.S.: Bioactive Sr(II)/chitosan/poly(ε-caprolactone) scaffolds for craniofacial tissue regeneration. In vitro and in vivo behavior. Polym. (Basel) **10**(3), 279 (2018)
42. Griffanti, G., Nazhat, S.N.: Dense fibrillar collagen-based hydrogels as functional osteoid-mimicking scaffolds. Int. Mater. Rev. **65**, 502–521 (2020)
43. Biswas, A., Bayer, I.S., Biris, A.S., Wang, T., Dervishi, E., Faupel, F.: Advances in top–down and bottom–up surface nanofabrication: techniques, applications & future prospects. Adv. Colloid Interfac. **170**, 2–27 (2012)
44. Wang, Z., Zhang, F., Wang, Z., Liu, Y., Fu, X., Jin, A., Yung, B.C., Chen, W., Fan, J., Yang, X., Niu, G., Chen, X.: Hierarchical assembly of bioactive amphiphilic molecule pairs into supramolecular nanofibril self-supportive scaffolds for stem cell differentiation. J. Am. Chem. Soc. **138**(45), 15027–15034 (2016)
45. Watcharajittanont, N., Putson, C., Pripatnanont, P., Meesane, J.: Electrospun polyurethane fibrous membranes of mimicked extracellular matrix for periodontal ligament: molecular behavior, mechanical properties, morphology, and osseointegration. J. Biomater. Appl. **34**, 753–762 (2020)
46. Wang, Z., Ustriyana, P., Chen, K., Zhao, W., Xu, Z., Sahai, N.: Toward the understanding of small protein-mediated collagen intrafibrillar mineralization. ACS Biomater. Sci. Eng. **6**(7), 4247–4255 (2020)
47. Ma, Y., Hoff, S.E., Huang, X., Liu, J., Wan, Q., Song, Q., Gu, J., Heinz, H., Tay, F.R., Niu, L.: Involvement of prenucleation clusters in calcium phosphate mineralization of collagen. Acta Biomater. **120**, 213–223 (2021)
48. Abbasi, N., Hamlet, S., Love, R.M., Nam-Nguyen, N.T.: Porous scaffolds for bone regeneration. J. Sci. Adv. Mater. Dev. **5**, 1–9 (2020)

49. Wang, X., Lou, T., Zhao, W., Song, G., Li, C., Cui, G.: The effect of fiber size and pore size on cell proliferation and infiltration in PLLA scaffolds on bone tissue engineering. J. Biomater. Appl. **30**, 1545–1551 (2016)
50. Chocholata, P., Kulda, V., Babuska, V.: Fabrication of Scaffolds for Bone-Tissue Regeneration. Materials (Basel) **12**(4), 568 (2019)
51. Wangkulangkul, P., Jaipaew, J., Puttawibul, P., Meesane, J.: Constructed silk fibroin scaffolds to mimic adipose tissue as engineered implantation materials in post-subcutaneous tumor removal. Mater. Design **106**, 428–435 (2016)
52. Jamalpoor, Z., Mirzadeh, H., Joghataei, M.T., Zeini, D., Bagheri-Khoulenjani, S., Nourani, M.R.: Fabrication of cancellous biomimetic chitosan-based nanocomposite scaffolds applying a combinational method for bone tissue engineering. J. Biomed. Mater. Res. A **103**, 1882–1892 (2015)
53. Kankala, R.K., Xu, X.M., Liu, G.G., Chen, A.Z., Wang, S.B.: 3D-printing of microfibrous porous scaffolds based on hybrid approaches for bone tissue engineering. Polym. (Basel) **10**(7), 807 (2018)
54. Fuchs, A., Youssef, A., Seher, A., Hartmann, S., Brands, R.C., Müller-Richter, U.D.A., Kübler, A.C., Linz, C.: A new multilayered membrane for tissue engineering of oral hard- and soft tissue by means of melt electrospinning writing and film casting – An in vitro study. J. Cranio Maxill. Surg. **47**, 695–703 (2019)
55. Glim, J.E., Everts, V., Niessen, F.B., Ulrich, M.M., Beelen, R.H.J.: Extracellular matrix components of oral mucosa differ from skin and resemble that of foetal skin. Arch. Oral Biol. **59**, 1048–1055 (2014)
56. Taufalele, P.V., Burgh, J.A.V., Muñoz, A., Zanotelli, M.R., Reinhart-King, C.A.: Fiber alignment drives changes in architectural and mechanical features in collagen matrices. PLoS ONE **14**(5), e0216537 (2019)
57. Groeger, S., Meyle, J.: Oral mucosal epithelial cells. Front. Immunol. **10**, 208 (2019)
58. Lu, Q., Wang, X., Lu, S., Li, M., Kaplan, D.L., Zhu, H.: Nanofibrous architecture of silk fibroin scaffolds prepared with a mild self-assembly process. Biomaterials **32**, 1059–1067 (2011)
59. Sangkert, S., Kamolmatyakul, S., Gelinsky, M., Meesane, J.: 3D printed scaffolds of alginate/polyvinylalcohol with silk fibroin based on mimicked extracellular matrix for bone tissue engineering in maxillofacial surgery. Mater. Today Commun. **26**, 102140 (2021)
60. Sant, S., Coutinho, D.F., Gaharwar, A.K., Neves, N.M., Reis, R.L., Gomes, M.E., Khademhosseini, A.: Self-assembled hydrogel fiber bundles from oppositely charged polyelectrolytes mimic micro-/nanoscale hierarchy of collagen. Adv. Funct. Mater. **27**, 1606273 (2017)
61. Huang, X., Bai, S., Lu, Q., Liu, X., Liu, S., Zhu, H.: Osteoinductive-nanoscaled silk/HA composite scaffolds for bone tissue engineering application. J. Biomed. Mater. Res. B **103**, 1402–1414 (2015)
62. Gassling, V., Douglas, T., Warnke, P.H., Açil, Y., Wiltfang, J., Becker, S.T.: Platelet-rich fibrin membranes as scaffolds for periosteal tissue engineering. Clin. Oral Implan. Res. **21**, 543–549 (2010)
63. Kesting, M.R., Wolff, K.D., Nobis, C.P., Rohleder, N.H.: Amniotic membrane in oral and maxillofacial surgery. Oral Maxillofac. Surg. **18**, 153–164 (2014)
64. Laurent, R., Nallet, A., de Billy, B., Obert, L., Nicod, L., Meyer, C., Layrolle, P., Zwetyenga, N., Gindraux, F.: Fresh and in vitro osteodifferentiated human amniotic membrane, alone or associated with an additional scaffold, does not induce ectopic bone formation in Balb/c mice. Cell Tissue Bank **18**, 17–25 (2017)
65. Schuckert, K.H., Jopp, S., Osadnik, M.: The use of platelet rich plasma, bone morphogenetic protein-2 and different scaffolds in oral and maxillofacial surgery - literature review in comparison with own clinical experience. J. Oral Maxillofac. Res. **2**(1), e2 (2011)
66. de Freitas, R.M., Spin-Neto, R., Junior, E.M., Pereira, L.A.V.D., Wikesjö, U.M.E., Susin, C.: Alveolar ridge and maxillary sinus augmentation using rhbmp-2: a systematic review. Clin. Orthop. Relat. Res. **17**, e192–e201 (2015)
67. Kwak, E.A., Lee, N.Y.: Synergetic roles of TGF-β signaling in tissue engineering. Cytokine **115**, 60–63 (2019)

68. Khojasteh, A., Fahimipour, F., Eslaminejad, M.B., Jafarian, M., Jahangir, S., Bastami, F., Tahriri, M., Karkhaneh, A., Tayebi, L.: Development of PLGA-coated β-TCP scaffolds containing VEGF for bone tissue engineering. Mat. Sci. Eng. C Mater. **69**, 780–788 (2016)
69. Jia, J., Siheng, W., Fang, C., Chengchong, A., Shiyi, C.: The study on vascularisation and osteogenesis of BMP/VEGF co-modified tissue engineering bone in vivo. RSC Adv. **6**, 41800–41808 (2016)
70. Soriente, A., Fasolino, I., Raucci, M.G., Demitri, C., Madaghiele, M., Giuri, A., Sannino, A., Ambrosio, L.: Effect of inorganic and organic bioactive signals decoration on the biological performance of chitosan scaffolds for bone tissue engineering. J. Mater. Sci. Mater. Med. **29**, 62 (2018)
71. Shi, H., Zhou, Z., Li, W., Fan, Y., Li, Z., Wei, J.: Hydroxyapatite based materials for bone tissue engineering: a brief and comprehensive introduction. Cryst. **11**, 149 (2021)
72. Yazdimamaghani, M., Razavi, M., Vashae, D., Moharamzadeh, K., Boccaccini, A.R., Tayebi, L.: Porous magnesium-based scaffolds for tissue engineering. Mater. Sci. Eng. C Mater. **71**, 1253–1266 (2017)
73. Lei, Y., Xu, Z., Ke, Q., Yin, W., Chen, Y., Zhang, C., Guo, Y.: Strontium hydroxyapatite/chitosan nanohybrid scaffolds with enhanced osteoinductivity for bone tissue engineering. Mater. Sci. Eng. C Mater. **72**, 134–142 (2017)
74. Kołodziejska, B., Stępień, N., Kolmas, J.: The influence of strontium on bone tissue metabolism and its application in osteoporosis treatment. Int. J. Mol. Sci. **22**(12), 6564 (2021)
75. Bose, S., Tarafder, S.: Calcium phosphate ceramic systems in growth factor and drug delivery for bone tissue engineering: a review. Acta Biomater. **8**, 1401–1421 (2012)

# Chapter 10
# Mimicked Hydrogel Scaffolds for Articular Cartilage Surgery

There are many patients who suffer from cartilage tissue defects, particularly due to disease and trauma. Generally, the soft tissue of cartilage is damaged from degeneration and/or accidents. In the case of damaged cartilage tissue into the bone phase, patients will have serious pain. In such a case, the damaged areas may need total replacement. In some cases, a tissue engineering scaffold is suitable as a substitute biomaterial to regenerate new tissue [1]. Creating suitable scaffolds to be similar to native tissue is important to induce tissue regeneration at the defect site. For some defects, in cartilage tissue, hydrogel scaffolds are selected for use in surgery. Hence, this chapter discusses performance hydrogel scaffolds, which are created based on the mimic approach. The geometry, morphology, structure, and functionality of performance scaffolds are considered in this chapter.

## 10.1 Hydrogel Scaffolds for Articular Cartilage Surgery

Articular cartilage is one part of the knee joint (Fig. 1a). Articular cartilage has the important role to receive and resist biomechanic forces during the movement of the human body. Articular cartilage tissue shows a structure of multilayer, tough tissue that comprises of tide make, deep, transitional, and superficial zones (Fig. 1b). Normally, the deep zone demonstrates the highest thickness. In each zone, there are chondrocytes embedded in the extracellular matrix (ECM), of which the main components are hyaluronic acid and collagen type II. Cartilage ECMs have high water content in their structure. This leads to the durable toughness, which provides for suitable mechanical performance for compressive forces from body weight.

Due to the articular cartilage obtaining high force loading, it is under risk conditions for tissue damage, particularly in patients who have problems from of degenerative diseases. Osteoarthritis is one such critical disease that often occurs in elderly patients. The pathology of osteoarthritis involves cartilage tissue degeneration that



J. Meesane, *Mimicked Tissue Engineering Scaffolds for Maxillofacial and Articular Cartilage Surgery*, Engineering Materials,
https://doi.org/10.1007/978-981-19-7830-2_10

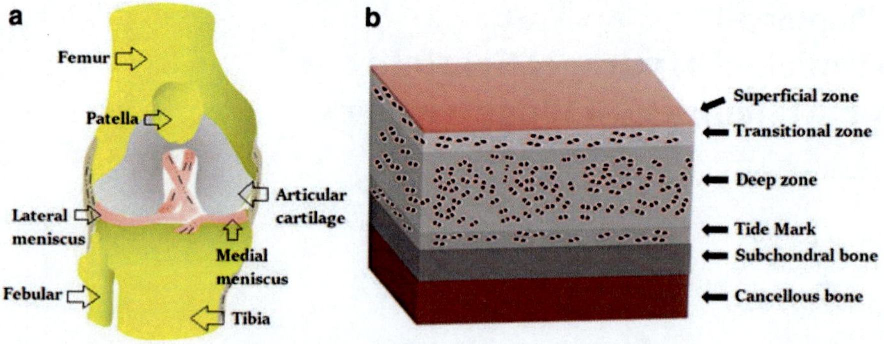


**Fig. 10.1** **a** Structure of the knee joint and **b** structure of articular cartilage

leads to a crucial defect area, this being the cause of pain during movement (Fig. 2a). In osteoarthritis, the articular cartilage has a thinner structure than normal cartilage tissue (Fig. 2b), and this thinner articular cartilage is at risk of developing microcracks. In turn, these microcracks lead to crucial defects in osteoarthritis, causing an irregular alignment of the legs. In critical cases, patients require an operation for partial or total knee replacement (Fig. 3a, b).

In some cases, hydrogel scaffolds are selected as performance biomaterials to regenerate new tissue at the defect site [2, 3]. Hydrogels show good performance for high water content, which is similar to cartilage tissue. Furthermore, some developed hydrogels have the function which matches to the performance of cartilage tissue, particularly in high compressive resistance. For surgery in articular cartilage, hydrogels can be used via two options: injectable and non-injectable. In the case of injectable hydrogel, the patient is operated on via non-invasive surgery (Fig. 4a). For this surgery, the solution of precursor is injected into the defect area, before being activated to form hydrogels. On the other hand, for non-injectable hydrogel, this is fabricated into the matched shape of the defect area, and then, it is inserted into the defect area (Fig. 4b).

For the requirements of hydrogel scaffolds, they are divided into two views: (1) the tissue engineering view and (2) the surgical view. In the view of tissue engineering, structure and function of hydrogel scaffolds are important requirements. For nano-or molecular structures, hydrogels have to show hydrophilic segments in their molecules, for instance—OH, NH3, CONH3, and COOH, as this is the cause of water uptake in hydrogel scaffolds. These hydrophilic segments show performance as the binding site of cells. This leads to promote cell adhesion and proliferation, which contribute to enhance tissue formation. Furthermore, the hydrophilic segment is the cause of hydrolytic degradation during tissue formation. This is a necessary factor to focus on for hydrogel scaffolds, because they have to demonstrate a suitable degradation rate matching that of new tissue formation, without collapse. For micro-or morphological structures, hydrogel needs suitable pore sizes for chondrocytes. Furthermore, these pores have to show an inter-connective structure, which

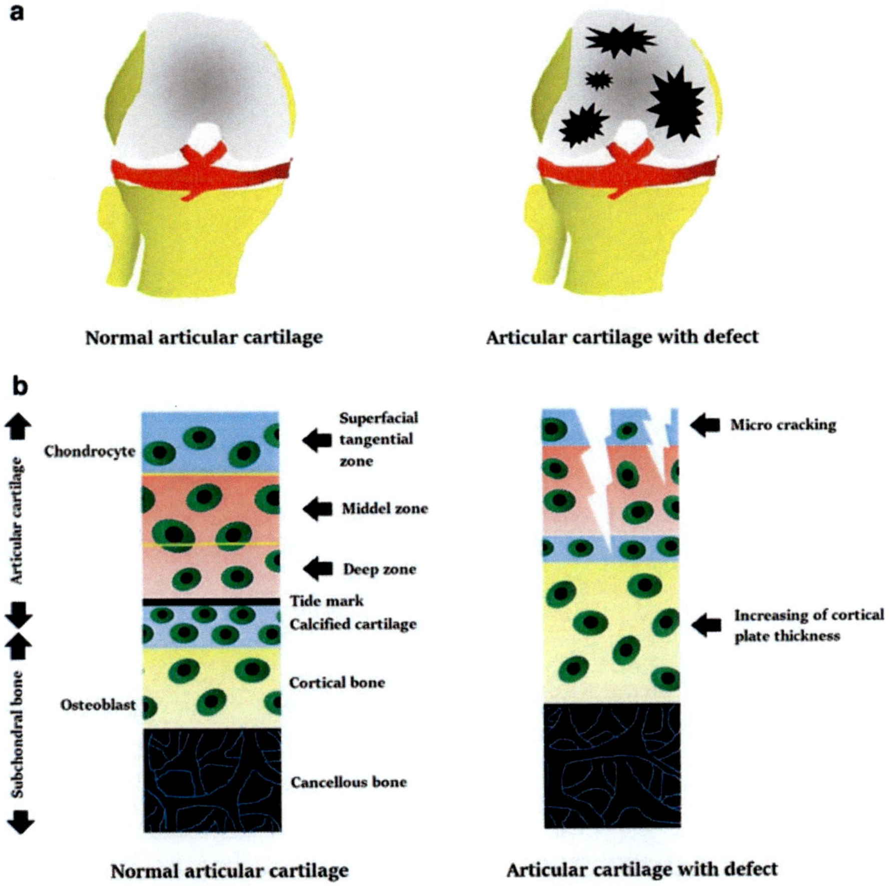


**Fig. 10.2** **a** Normal articular cartilage and with defect and **b** morphological structure of normal articular cartilage and with defect

contributes to nutrient and biological signals of diffusion. This assists to promote cartilage tissue formation.

For function, hydrogel is required to have physical, mechanical, and biological properties which support new tissue formation. Water uptake is an important physical property, which is reflected in the performance of hydrogel scaffolds. Hydrogel scaffolds have to show water uptake that contributes to the nutrient and biological signals of diffusion, so as to promote new tissue formation. Additionally, the water uptake of hydrogel is the cause of swelling behaviors. In the case of high swelling behaviors, hydrogel scaffolds show high, expanded sizes, which are non-fitting to the defect site. Therefore, good hydrogel scaffolds are required to have suitable swelling behaviors which fit to the defect site.

Mechanical strength is another important mechanical property required for hydrogel scaffolds. To avoid the initiation of defect at interface, hydrogel scaffolds

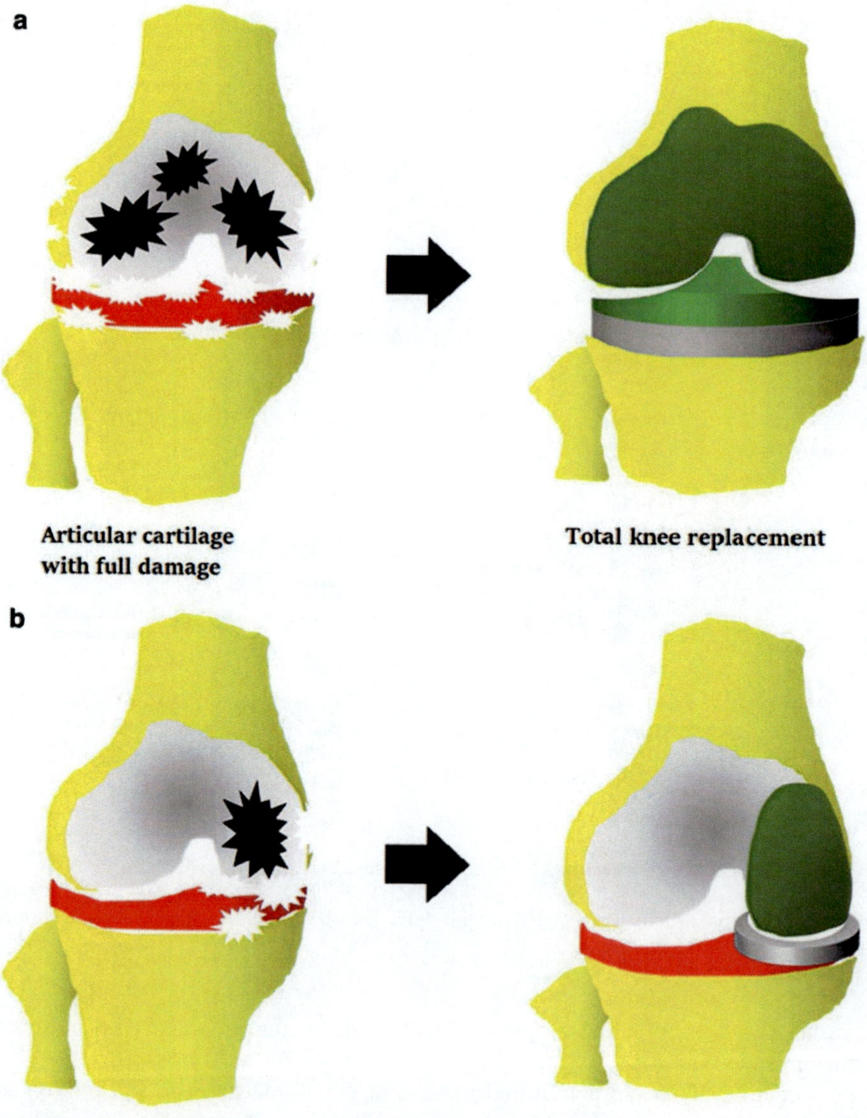


**Fig. 10.3** **a** Articular cartilage with full damage and total knee replacement and **b** articular cartilage with partial damage and partial knee replacement

have to show compressive strength matching that of cartilage tissue. If the defect leads to a collapse of the hydrogel scaffold during tissue formation, it will have an effect on irregular formation of new tissue.

Biocompatibility and biodegradation are important biological properties of hydrogel scaffolds. For biocompatibility, hydrogel scaffolds have to promote cell adhesion and proliferation, which leads to the enhancement of new tissue formations.

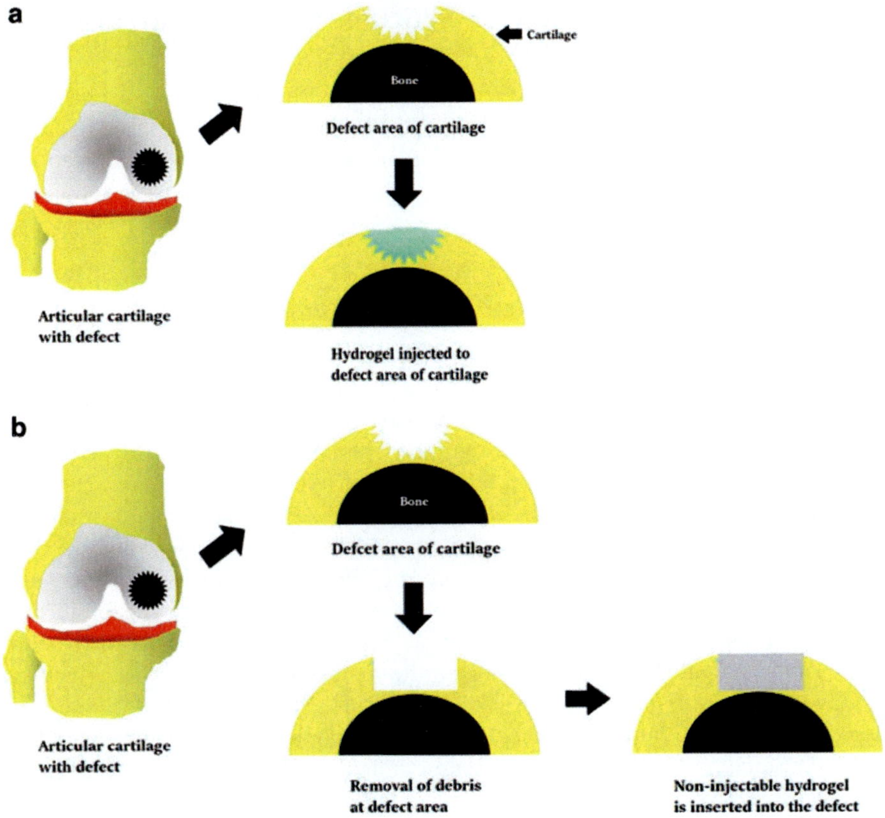


**Fig. 10.4** **a** Surgery of articular cartilage with injectable hydrogel and **b** non-injectable hydrogel

Some smart hydrogels, incorporated with stem cells, are injected into the defect area of the cartilage [4]. Hydrogels have the unique functionality to form a gel at the defect site, under the right pH, thermal, or ion sensitivity. Nevertheless, some hydrogels show insufficient stability. This leads to their removal from the defect area to the outside environment of cartilage tissue. This leads to the irregular formation of new tissue. To solve this problem, hydrogel is incorporated with 2D membrane scaffolds, which are covered on the surface of the defect [5], and this prevents the leakage of cells. Details of 2D scaffolds for cartilage surgery are further described in Chapter 12. A scaffold is designed into biphasic 3D scaffolds of barrier membranes, affixed on a sponge material [6]. Barrier membranes are used to prevent cell leakage, and the sponge material acts as a structure for promotion of cartilage tissue formation [5]. Importantly, in the creation of the scaffold hydrogel, with similar structure and function as that of the microenvironment of the extracellular matrix (ECM), growth factors or cytokine is the potential approach to induce new cartilage tissue formation without malfunction [7].

## 10.2 Mimicked Hydrogels for Articular Cartilage Surgery

Mimicking is the approach used to create hydrogel scaffolds similar to the microenvironment of tissue. Especially in the view of tissue engineering, mimicked hydrogel scaffolds are designed with similar structural and functional microenvironments of cartilage. This leads to their high potential to induce new cartilage tissue formation.

### *10.2.1 Mimicked Structure of Microenvironments for Hydrogel*

For mimicking the structure of the microenvironment, three levels of structure: (1) molecular or nano-structure, (2) morphological or microstructure, and (3) geographical or macro-structures are used in the design for hydrogel scaffolds in cartilage tissue [8–10]. For their molecular structure, polypeptide with amino acid, similar to collagen type II which is the main protein component in cartilage, is synthesized by mimicking of the whole molecular structure [11]. For some designs, the amino acid sequence, which acts as the cell binding site, is synthesized based on the mimicking of a partial molecular structure [12, 13]. Bone morphogenetic proteins (BMPs) and transforming growth factor-β (TGF-β) are the main growth factors in the role of inducing cartilage tissue formation [14, 15]. For this first approach, to mimic the molecular structure, they are synthesized with the whole molecule. For the second approach, some parts of growth factors are mimicked for a partial molecular structure. They are produced with protein engineering or bioprocessing of recombinant protein [12, 16, 17].

In the case of morphological structures, fibril of collagen is mimicked for hydrogel scaffolds. For instance, collagen type II is selected as the base material for mimicking of fibril in cartilage tissue [18]. Fibrils of collagen type II have self-organization of fibril network structure, which are suitable for chondrocyte behaviors, leading to enhancement of cartilage tissue formations [19, 20]. Pore size and structure are mimicked for fitting to adhesion and self-organization in spherical shapes of chondrocyte [21, 22]. Furthermore, the size and structure of the pores have to contribute to the diffusion of nutrients, biological signals, and cytokines that promote tissue formation [23, 24].

To mimic geographical structures, hydrogel is connected to 3D sponges, to construct biphasic scaffolds, which shows similar anatomy of cartilage [25]. For this biphasic scaffold, hydrogel is used as cartilage tissue. On the other hand, the 3D sponge is proposed for bone tissue.

### *10.2.2 Mimicked Function of Microenvironments for Hydrogel*

For the mimicking in the function of a microenvironment, physical, mechanical, and biological performance is used for hydrogels [25, 26]. In the case of physical performance, swelling behaviors of this are evaluated for its potential of mimicked hydrogels [27]. The swelling behavior relates to performance of nutrients, growth factors, and cytokine diffusion, which is of importance for promoting the cartilage formation [28]. Normally, the microenvironment contains collagen type II and hyaluronic acid (HA) as the main ECMs, which have the role to promote swelling behavior [29]. Especially HA, which has a high molecular weight, negative charge, and hydrophilic performance, therefore has an effect on the high swelling behavior [30]. HA is often used as a base material to mimic hydrogel for cartilage [7, 31]. This is used as guidance in selecting the hydrophilic polymers that have a high molecular weight, so as to mimic the swelling behavior similar to HA [32, 33]. For instance, derivative cellulose, alginate, and chitosan are chosen to the mimic swelling behavior similar to HA [3, 34].

For mechanical function, mimicked hydrogel focuses on compressive behavior, this being the main mechanical performance obtained under biomechanic conditions of cartilage tissue [35]. To mimic compressive strength, similar to cartilage tissue, high molecular weight polymers are used to fabricate hydrogels which resist the obtained biomechanic force of cartilage [36, 37].

In the case of biological mimicking, natural or synthetic polymers, which have the functional groups to promote cell behaviors for enhancement tissue formation, are used as hydrogels [36]. Functional groups which demonstrate the hydrophilic and charge act as non-specific binding sites [38]. These binding sites demonstrate attachment with cells via physical interaction: electrostatic and hydrophilic interaction [39, 40]. For instance, chitosan molecules which have positive charges on their molecules have physical interaction with the negative charge on the surface of cells [41]. This has an effect on promotion of cell spreading, proliferation, and migration that enhance tissue formation [42]. On the other hand, some polymers, which have specific bindings site for cell recognition, are used to mimic hydrogels [42]. Examples of these are collagens and gelatin which have amino sequence of Arginine-Glycine-Aspartate (RGD) for cell recognition and demonstrate suitable performance to mimic hydrogels [43]. Furthermore, non-specific and specific binding site shows performance to interact with some growth factors or cytokine, which induce new tissue formation [44]. Nevertheless, to create mimicked hydrogels, which have potential to enhance new tissue formation, suitable fabrication and design must be undertaken.

## 10.3 Fabrication in Mimicked Hydrogels

Fabrication is of importance for mimicked hydrogels that have the potential to enhance new tissue formation. There are two main parts in the fabrication of hydrogel: (1) material selection and (2) approach of the hydrogel formation. Firstly, in the case of material selection, natural and synthetic polymers demonstrating biocompatibility and biodegradability have often been used to design mimicked hydrogels for cartilage tissue engineering. For natural polymers, hyaluronic (HA) is a popular performance hydrogel for cartilage tissue, because of its biocompatibility, biodegradability, and physical and mechanical properties.

HA is an important component of the extracellular matrix in cartilage tissue; therefore, HA is often selected for mimicked hydrogels. The structure of HA has long chain molecules containing carboxylic groups (–COOH), and these carboxylic groups enhance water adsorption. This results in compressive force resistance of cartilage tissue. Besides biological polymers as hydrogels for osteoarthritis surgery, synthetic polymers are also used. For synthetic polymers, polyvinyl (alcohol) (PVA) is an example of mimicked hydrogel, because of its biocompatibility, degradability, and processability. Examples of polymers for cartilage tissue engineering are shown in Table 10.1.

Secondly, the approach of hydrogel formation is shown as two types: (1) chemical and (2) physical crosslinking. Basically, for hydrogel formation, polymer molecules are connected to each other through chemical or physical crosslinking to build a network structure.

In the case of chemical crosslinking, molecules of crosslinkers, which have two reactive groups, are used to react with the molecules of the polymers (Fig. 10.5). Popular crosslinkers used for the formation of network structures include glutaraldehyde, 1-ethyl-3-(3-dimethylaminopropyl) carbodiimide hydrochloride, genipin, citric acid, and ascorbic acid.

Some polymers, which have reactive groups in their molecules, are selected for hydrogel formation via chemical crosslinking. These polymers are activated with sufficient energy: thermal, chemical, photo-irradiation and so forth. Then, these reactive groups form chemical crosslinking between polymer molecules. For

**Table 10.1** Examples of polymers for cartilage tissue engineering

| Biological polymers | Synthesis polymer |
|---|---|
| Hyaluronic acid (HA) [45]<br>Chitosan [46]<br>Chondroitin sulfate [47]<br>Collagen [48]<br>Gelatin [49]<br>Fibrin [50]<br>Silk fibroin [51]<br>Alginate [52]<br>Dextran [53]<br>Gellan [54, 55] | Polyvinyl alcohol (PVA) [56]<br>Polyethylene glycol (PEG) [57] |

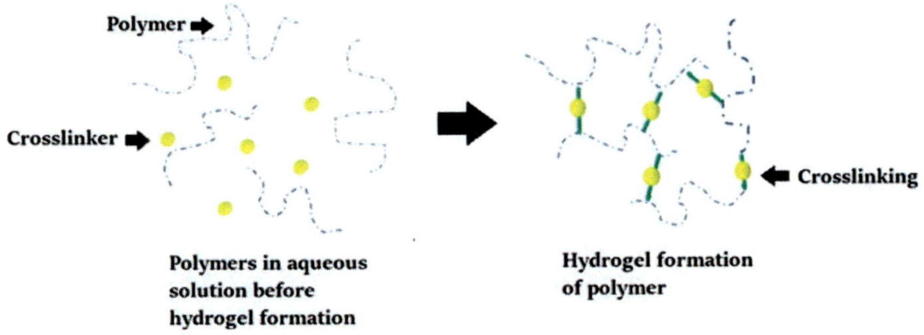


**Fig. 10.5** Formation of chemical crosslinking

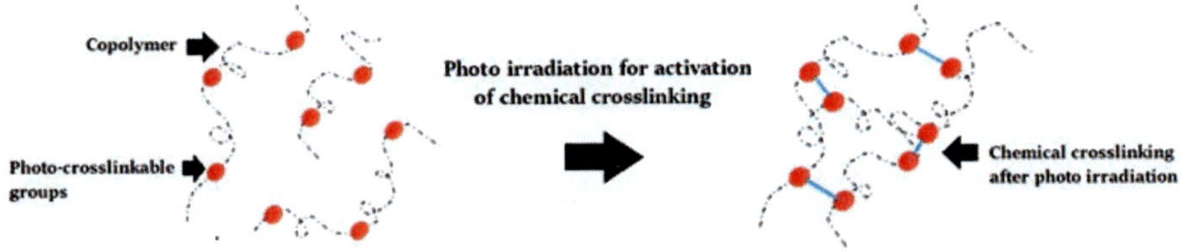


**Fig. 10.6** Mechanism of chemical reaction of a photo-reactive hydrogel

instance, polymers which are modified by grafting with photo-crosslinkable groups are activated by photo-irradiation to form chemical crosslinking [58] (Fig. 10.6).

In the case of physical crosslinking, this is formed via aggregation, electrostatic interaction, and crystal domain. For instance, poly (N-isopropylacrylamide) (PNIPAM) has often been used as a physical, crosslinked hydrogel [59]. At low temperatures, PNIPAM is in its solution state, and the molecules of PNIPAM are caged by water molecules. At normal, human body temperature, the solution transforms into a hydrogel. Then, hydrogel formation begins by physical interaction between the water and polymer molecules that are broken up by the higher temperature; these chemical groups are then free to form aggregations. This kind of hydrogel is called a thermo-sensitive hydrogel. The mechanism of PNIPAM thermo-sensitive hydrogel formation is shown in Fig. 10.7.

In the other instance, alginate shows crosslinking via electrostatic interaction of calcium ions, with a negative charge of carboxylic groups on its molecules [60]. PVA forms hydrogel with freeze thawing. Freeze thawed PVA has a crystal domain formation via its hydroxyl groups. This crystal domain acts as a physical crosslinking [61]. According to these examples of fabrication, they are important tools to design mimicked hydrogels that are fitted to target of application. Nevertheless, to further create better performance of mimicked hydrogel, modification must be undertaken.

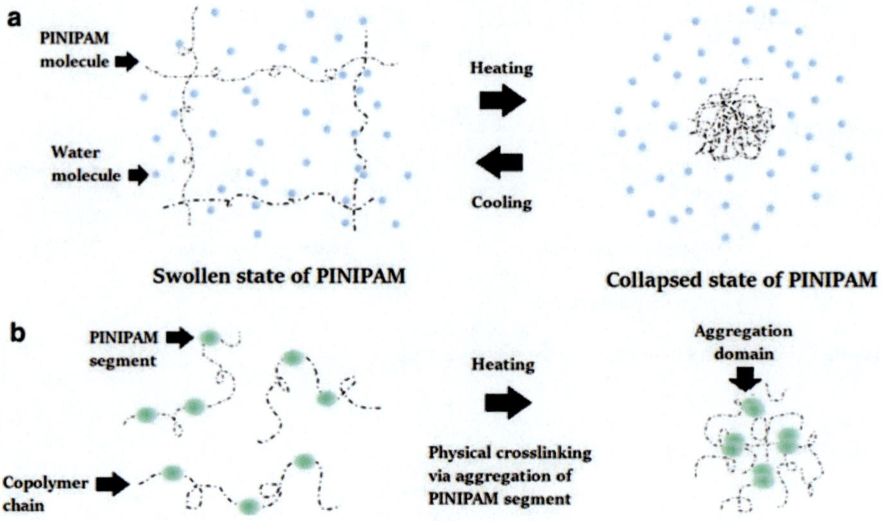


**Fig. 10.7** Mechanism of PNIPAM thermo-sensitive hydrogel: **a** thermo-sensitive behavior of PNIPAM and **b** thermo-sensitive behavior of PNIPAM copolymer

## 10.4 Modification in Mimicked Hydrogels

After fabrication, scaffolds show performance to support tissue formation. Modification is an important tool to create mimicked hydrogels that have high performance. Some hydrogels are modified with biological domains: degradable molecular segment and cell recognized motif, to promote high performance for inducing of tissue formation [62]. The conjugation of an enzyme degradable domain and incorporation of biological signals are the instance for the modification of mimicked hydrogels [63]. This is used to guide the creation of high performance mimicked hydrogels [64].

### *10.4.1 Conjugation of Enzyme Degradable Domains in Mimicked Hydrogels*

For enzyme degradation, it is a biological process during tissue formation. Normally, ECMs, which are structure for cell behaviors, are degraded by enzymes of the matrix metalloproteinase (MMPs). These enzymes attach to the specific binding domains of ECMs, upon which the ECMs are digested with MMP at these domains. This leads to the scission of ECMs and shows free spaces for cell proliferation and migration. Mimicked hydrogel, with enzyme degradable domains, is an approach to contribute to

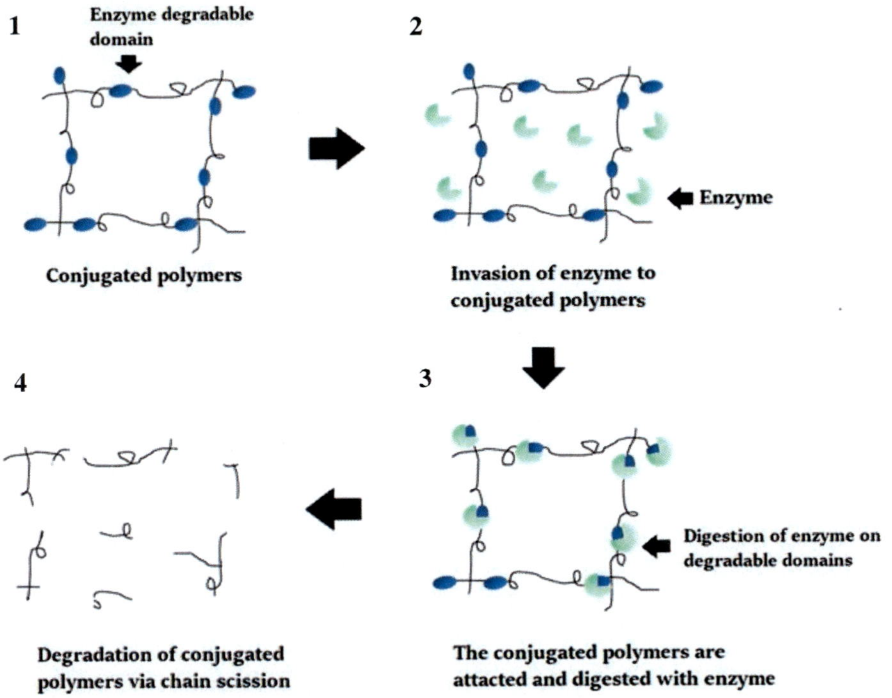


**Fig. 10.8** Enzyme degradation of conjugated polymers

tissue formation. For modification in mimicked hydrogels with these domains, polymers are normally conducted via conjugation. For the molecular structure, the conjugated polymers demonstrate enzyme degradable domains in their chains (Fig. 10.8). Degradation can occur when the conjugated polymers are digested by an enzyme [65]. First, enzymes invade the conjugated polymers, before attachment to the degradable domains, and these domains are then digested by enzymes. This leads to degradation via chain scission.

### *10.4.2 Incorporation of Biological Signals in Mimicked Hydrogels*

Besides the conjugated polymers with enzyme degradable domains that are coordinated with tissue regeneration, the biological functionality must bring about new tissue formation. Therefore, incorporation of biological signals in mimicked hydrogels has to have a specific functionality for new tissue formation; this is important in the design of the systems. Some researchers created these functionalities to be similar to a microenvironment that could enhance the performance for new tissue

**Table 10.2** Examples of extracellular matrix and growth factors in cartilage tissue

| Extracellular matrix | Growth factors |
|---|---|
| Proteoglycan type I<br>Aggrecan<br>Asporin<br>Biglycan<br>Chondroitin sulfate<br>Collagen type I<br>Collagen type II<br>Collagen IV<br>Collagen V<br>Collagen type VI<br>Collagen type IX<br>Collagen type XI<br>Decorin<br>Fibromodulin<br>Heparan sulfate<br>Hyaluronic acid (HA)<br>Matrilin 1<br>Matrilin 3 | Bone morphogenetic protein-2 (BMP-2)<br>Bone morphogenetic protein-7 (BMP-7)<br>Fibroblast growth factor-2 (FGF-2)<br>Fibroblast growth factor-18 (FGF-18)<br>Insulin-like growth factor-I (IGF-I)<br>Platelet-derived growth factor (PDGF)<br>Transforming growth factor-β1 (TGF-β1) |

formation [15]. The components of the microenvironment are the extracellular matrix and growth factors, and they are often added into hydrogels at specific ratios; it was found that the release profiles and bioactive activity were similar to native tissue [66]. Examples of extracellular matrix and growth factors in cartilage tissue are shown in Table 10.2 [67, 68].

For incorporation, physical and chemical methods are used in mimicked hydrogels [69]. After selection of suitable polymers and the approach of hydrogel formations, biological signals are chosen for incorporation. In the case of a physical method, mixing is often used to incorporate biological signals with mimicked hydrogel. For instance, alginate and physical crosslinking via electrostatic interaction are selected as base materials and approaches of hydrogel formations. Bone morphogenetic protein-2 (BMP-2) and transforming growth factor-β (TGF-β) or its encoding of genes are selected as base biological signals for mimicked hydrogel [70]. Alginate solution is mixed with BMP-2 and collagen type II, before the addition of CaCl2. This leads to hydrogel formation via electrostatic interaction between calcium cations with carboxylic anions in molecules of alginate. Biological signals of BMP-2 and TGF-β are then trapped in hydrogel.

For the chemical method, conjugation and grafting are used to incorporate biological signals in mimicked hydrogel [71]. These methods show chemical bonding in the polymer molecules. For instance, multiarm polyethylene glycol (multiarm PEG) is selected as a base polymer, which then forms hydrogel via chemical crosslinking. RGD or YIGSR are the amino acid motif, which are selected as biological signals. This PEG is incorporated with biological signals via conjugation. Then, the biological signal conjugated PEG is added as crosslinkers to form hydrogel via chemical bonding [60]. Based on these instances, regarding both physical and chemical methods, mimicked hydrogels can be created with varied performances, depending on the aim of application in cartilage tissue engineering.

## References

1. Wasyłeczko, M., Sikorska, W., Chwojnowski, A.: Review of synthetic and hybrid scaffolds in cartilage tissue engineering. Membr. (Basel) **10**(11), 348 (2020)
2. Li, J., Chen, G., Xu, X., Abdou, P., Jiang, Q., Shi, D., Gu, Z.: Advances of injectable hydrogel-based scaffolds for cartilage regeneration. Regen. Biomater. **6**(3), 129–140 (2019)
3. Li, L., Yu, F., Zheng, L., Wang, R., Yan, W., Wang, Z., Xu, J., Wu, J., Shi, D., Zhu, L., Wang, X., Jiang, Q.: Natural hydrogels for cartilage regeneration: modification, preparation and application. J. Orthop. Translat. **14**(17), 26–41 (2018)
4. Liu, M., Zeng, X., Ma, C., Yi, H., Ali, Z., Mou, X., Li, S., Deng, Y., He, N.: Injectable hydrogels for cartilage and bone tissue engineering. Bone Res. **5**, 17014 (2017)
5. Makris, E.A., Gomoll, A.H., Malizos, K.N., Hu, J.C., Athanasiou, K.A.: Repair and tissue engineering techniques for articular cartilage. Nat. Rev. Rheumatol. **11**, 21–34 (2015)
6. Panjapheree, K., Kamonmattayakul, S., Meesane, J.: Biphasic scaffolds of silk fibroin film affixed to silk fibroin/chitosan sponge based on surgical design for cartilage defect in osteoarthritis. Mater. Design **141**, 323–332 (2018)
7. Jaipaew, J., Wangkulangkul, P., Meesane, J., Raungrut, P., Puttawibul, P.: Mimicked cartilage scaffolds of silk fibroin/hyaluronic acid with stem cells for osteoarthritis surgery: morphological, mechanical, and physical clues. Mater. Sci. Eng. C Mater. **64**, 173–182 (2016)
8. Ngadimin, K.D., Stokes, A., Gentile, P., Ferreira, A.M.: Biomimetic hydrogels designed for cartilage tissue engineering. Biomater. Sci. **9**, 4246–4259 (2021)
9. Li, Y., Liu, Y., Xun, X., Zhang, W., Xu, Y., Gu, D.: Three-dimensional porous scaffolds with biomimetic microarchitecture and bioactivity for cartilage tissue engineering. ACS Appl. Mater. Interfaces **11**(40), 36359–36370 (2019)
10. Fu, L., Yang, Z., Gao, C., Li, H., Yuan, Z., Wang, F., Sui, X., Liu, S., Guo, Q.: Advances and prospects in biomimetic multilayered scaffolds for articular cartilage regeneration. Regen. Biomater. **7**, 527–542 (2020)
11. Irawan, V., Sung, T.C., Higuchi, A., Ikoma, T.: Collagen scaffolds in cartilage tissue engineering and relevant approaches for future development. Tissue Eng. Regen. Med. **15**(6), 673–697 (2018)
12. Park, K.M., Joung, Y.K., Park, K.D., Lee, S.Y., Myung Lee, M.C.: RGD-Conjugated chitosan-pluronic hydrogels as a cell supported scaffold for articular cartilage regeneration. Macromol. Res. **16**, 517–523 (2008)
13. Chen, F., Ni, Y., Liu, B., Zhou, T., Yu, C., Su, Y., Zhu, X., Yu, X., Zhou, Y.: Self-crosslinking and injectable hyaluronic acid/RGD-functionalized pectin hydrogel for cartilage tissue engineering. Carbohyd. Polym. **166**, 31–44 (2017)
14. Madry, H., Rey-Rico, A., Venkatesan, J.K., Johnstone, B., Cucchiarini, M.: Transforming growth factor beta-releasing scaffolds for cartilage tissue engineering. Tissue Eng. Pt B Rev. **20**, 106–125 (2014)
15. Chen, L., Liu, J., Guan, M., Zhou, T., Duan, X., Xiang, Z.: Growth factor and its polymer scaffold-based delivery system for cartilage tissue engineering. Int. J. Nanomed. **15**, 6097–6111 (2020)
16. Overton, T.W.: Recombinant protein production in bacterial hosts. DrugRUG Discov. Today **19**, 590–601 (2014)
17. Poudel, S.B., Min, C.K., Lee, J.H., Shin, Y.J., Kwon, T.H., Jeon, Y.M., Lee, J.C.: Local supplementation with plant-derived recombinant human FGF2 protein enhances bone formation in critical-sized calvarial defects. J. Bone Miner. Metab. **37**, 900–912 (2019)

18. Almeida, H.V., Sathy, B.N., Dudurych, I., Buckley, C.T., O'Brien, F.J., Kelly, D.J.: Anisotropic shape-memory alginate scaffolds functionalized with either Type I or Type II collagen for cartilage tissue engineering. Tissue Eng. Pt A **23**, 55–68 (2017)
19. Liu, X., Zheng, C., Luo, X., Wang, X., Jiang, H.: Recent advances of collagen-based biomaterials: Multi-hierarchical structure, modification and biomedical applications. Mater. Sci. Eng. C Mater. **99**, 1509–1522 (2019)
20. Lee, C.R., Breinan, H.A., Nehrer, S., Spector, M.: Articular cartilage chondrocytes in Type I and Type II collagen-GAG matrices exhibit contractile behavior in vitro. Tissue Eng. Pt A **6**, 555–565 (2000)
21. Woodfield, T.B.F., Blitterswijk, C.A., Wijn, J., Sims, T.J., Hollander, A.P., Riesle, J.: Polymer scaffolds fabricated with pore-size gradients as a model for studying the zonal organization within tissue-engineered cartilage constructs. Tissue Eng. **11**(9–10), 1297–1311 (2005)
22. Zhang, Q., Lu, H., Kawazo, N., Chen, G.: Pore size effect of collagen scaffolds on cartilage regeneration. Acta Biomater. **10**, 2005–2013 (2014)
23. Karande, T.S., Ong, J.L., Agrawal, C.M.: Diffusion in musculoskeletal tissue engineering scaffolds: design issues related to porosity, permeability, architecture, and nutrient mixing. Ann. Biomed. Eng. **32**, 1728–1743 (2004)
24. Griffon, D.J., Sedighi, M.R., Schaeffer, D.V., Eurell, J.A., Johnson, A.L.: Chitosan scaffolds: interconnective pore size and cartilage engineering. Acta Biomater. **2**, 313–320 (2006)
25. Zheng, X.F., Lu, S.B., Zhang, W.G., Liu, S.Y., Huang, J.X., Guo, Q.Y.: Mesenchymal stem cells on a decellularized cartilage matrix for cartilage tissue engineering. Biotechnol. Bioproc. E **16**, 593–602 (2011)
26. Burdick, J.A., Vunjak-Novakovic, G.: Engineered microenvironments for controlled stem cell differentiation. Tissue Eng. Pt A **15**, 205–219 (2009)
27. Thangprasert, A., Tansakul, C., Thuaksubun, N., Meesane, J.: Mimicked hybrid hydrogel based on gelatin/PVA for tissue engineering in subchondral bone interface for osteoarthritis surgery. Mater. Design **183**, 108113 (2019)
28. Ahn, G., Kim, Y., Lee, S.W., Jeong, Y.J., Son, H., Lee, D.: Effect of heterogeneous multi-layered gelatin scaffolds on the diffusion characteristics and cellular activities of preosteoblasts. Macromol. Res. **22**, 99–107 (2014)
29. Guo, Y., Yuan, T., Xiao, Z., Tang, P., Xiao, Y., Fan, Y., Zhang, X.: Hydrogels of collagen/chondroitin sulfate/hyaluronan interpenetrating polymer network for cartilage tissue engineering. J. Mater. Sci. Mater. Med. **23**, 2267–2279 (2012)
30. Servaty, R., Schiller, J., Binder, H., Arnold, K.: Hydration of polymeric components of cartilage — an infrared spectroscopic study on hyaluronic acid and chondroitin sulfate. Int. J. Biol. Macromol. **28**, 121–127 (2001)
31. Kim, I.L., Mauck, R.L., Burdick, J.A.: Hydrogel design for cartilage tissue engineering: a case study with hyaluronic acid. Biomaterials **32**, 8771–8782 (2011)
32. Eslahi, N., Abdorahim, M., Simchi, A.: Smart polymeric hydrogels for cartilage tissue engineering: a review on the chemistry and biological functions. Biomacromol. **17**(11), 3441–3463 (2016)
33. Zhang, Y., Yu, J., Ren, K., Zuo, J., Ding, J., Chen, X.: Thermosensitive hydrogels as scaffolds for cartilage tissue engineering. Biomacromol. **20**(4), 1478–1492 (2019)
34. Jin, R., Moreira Teixeira, L.S., Dijkstra, P.J., Karperien, M., van Blitterswijk, C.A., Zhong, Z.Y., Feijen, J.: Injectable chitosan-based hydrogels for cartilage tissue engineering. Biomaterials **30**(13), 2544–2551 (2009)
35. Martínez-Moreno, D., Jiménez, G., Gálvez-Martín, P., Rus, G., Marchal, J.A.: Cartilage biomechanics: a key factor for osteoarthritis regenerative medicine. BBA Mol. Basis Dis. **1865**, 1067–1075 (2019)
36. Hafezi, M., Nouri Khorasani, S., Zare, M., Esmaeely Neisiany, R., Davoodi, P.: Advanced hydrogels for cartilage tissue engineering: recent progress and future directions. Polym. Basel **13**, 4199 (2021)
37. Mazzoccoli, J.P., Feke, D.L., Baskaran, H., Pintauro, P.N.: Mechanical and cell viability properties of crosslinked low-and high-molecular weight poly(ethylene glycol) diacrylate blends. J. Biomed. Mater. A **93A**, 558–566 (2010)

38. Sterner, B., Harms, M., Wöll, S., Weigandt, M., Windbergs, M., Lehr, C.M.: The effect of polymer size and charge of molecules on permeation through synovial membrane and accumulation in hyaline articular cartilage. Eur. Pharm. Biopharm. **101**, 126–136 (2016)
39. Hoshiba, T., Yoshikawa, C., Sakakibara, K.: Characterization of initial cell adhesion on charged polymer substrates in serum-containing and serum-free media. Langmuir **34**(13), 4043–4051 (2018)
40. Benhabbour, S.R., Sheardown, H., Adronov, A.: Cell adhesion and proliferation on hydrophilic dendritically modified surfaces. Biomaterial **29**, 4177–4186 (2008)
41. Carvalho, C.R., López-Cebral, R., Silva-Correia, J., Silva, J.M., Mano, J.F., Silva, T.H., Freier, T., Reis, R.L., Oliveira, J.M.: Investigation of cell adhesion in chitosan membranes for peripheral nerve regeneration. Mater. Sci. Eng. C Mater. **71**, 1122–1134 (2017)
42. Li, H., Hu, C., Yu, H., Chen, C.: Chitosan composite scaffolds for articular cartilage defect repair: a review. RSC Adv. **8**, 3736–3749 (2018)
43. Bellis, S.L.: Advantages of RGD peptides for directing cell association with biomaterials. Biomaterials **32**(18), 4205–4210 (2011)
44. Tjong, W.Y., Lin, H.H.: The RGD motif is involved in CD97/ADGRE5-promoted cell adhesion and viability of HT1080 cells. Sci. Rep. **9**, 1517 (2019)
45. Park, S.H., Seo, J.Y., Park, J.Y., Ji, Y.B., Kim, Y., Choi, H.S., Choi, S., Kim, J.H., Min, B.H., Kim, M.S.: An injectable, click-crosslinked, cytomodulin-modified hyaluronic acid hydrogel for cartilage tissue engineering. NPG Asia Mater **11**, 30 (2019)
46. Dehghan-Baniani, D., Chen, Y., Wang, D., Bagheri, R., Solouk, A., Wu, H.: Injectable in situ forming kartogenin-loaded chitosan hydrogel with tunable rheological properties for cartilage tissue engineering. Colloid Surface B **192**, 111059 (2020)
47. Li, X., Xu, Q., Johnson, M., Wang, X., Lyu, J., Li, Y., McMahon, S., Greiser, U., Sigen, A., Wang, W.: A chondroitin sulfate based injectable hydrogel for delivery of stem cells in cartilage regeneration. Biomater. Sci. **9**, 4139–4148 (2021)
48. Kilmer, C.E., Battistoni, C.M., Cox, A., Breur, G.J., Panitch, A., Liu, J.C.: Collagen Type I and II blend hydrogel with autologous mesenchymal stem cells as a scaffold for articular cartilage defect repair. ACS Biomater. Sci. Eng. **6**(6), 3464–3476 (2020)
49. Wang, K.Y., Jin, X.Y., Ma, Y.H., Cai, W.J., Xiao, W.Y., Li, Z.W., Qi, X., Ding, J.: Injectable stress relaxation gelatin-based hydrogels with positive surface charge for adsorption of aggrecan and facile cartilage tissue regeneration. J. Nanobiotechnol. **19**, 214 (2021)
50. Gupta, N., Cruz, M.A., Nasser, P., Rosenberg, J.D., Iatridis, J.C.: Fibrin-genipin hydrogel for cartilage tissue engineering in nasal reconstruction. Ann. Otol. Rhinol. Laryngol. **128**(7), 640–646 (2019)
51. Yua, T., Li, Z., Zhang, Y., Shen, K., Zhang, X., Xie, R., Liu, F., Fan, W.: Injectable Ultrasonication-Induced Silk Fibroin Hydrogel for Cartilage Repair and Regeneration. Tissue Eng. Pt. A **27**(17–18), 1213–1224 (2021)
52. Shojarazavi, N., Shojarazavi, N., Mashayekhan, S., Pazooki, H., Mohsenifard, S., Baniasadi, H.: Alginate/cartilage extracellular matrix-based injectable interpenetrating polymer network hydrogel for cartilage tissue engineering. J. Biomater. Appl. **36**(5), 803–817 (2021)
53. Wang, X., Li, Z., Shi, T., Zhao, P., An, K., Lin, C., Liu, H.: Injectable dextran hydrogels fabricated by metal-free click chemistry for cartilage tissue engineering. Mater. Sci. Eng. C Mater. **73**, 21–30 (2017)
54. Chen, Q., Zhang, D., Zhang, W., Zhang, H., Zou, J., Chen, M., Li, J., Yuan, Y., Liu, R.: Dual mechanism β-amino acid polymers promoting cell adhesion. Nat. Commun. **12**(562) (2021)
55. Kim, W., Choi, J.H., Kim, P., Youn, J., Song, J.E., Motta, A., Migliaresi, C., Khang, G.: Preparation and evaluation of gellan gum hydrogel reinforced with silk fibers with enhanced mechanical and biological properties for cartilage tissue engineering. J. Tissue Eng. Regen. Med. **15**(11), 936–947 (2021)
56. Nazouri, M., Seifzadeh, A., Masaeli, E.: Characterization of polyvinyl alcohol hydrogels as tissue-engineered cartilage scaffolds using a coupled finite element-optimization algorithm. J. Biomech. **99**, 109525 (2020)

57. Rennerfeldt, D.A., Renth, A.N., Talata, Z., Gehrke, S.H., Detamore, M.S.: Tuning mechanical performance of poly(ethylene glycol) and agarose interpenetrating network hydrogels for cartilage tissue engineering. Biomaterials **34**(33), 8241–8257 (2013)
58. Wang, Y., Koole, L.H., Gao, C., Yang, D., Yang, L., Zhang, C., Li, H.: The potential utility of hybrid photo-crosslinked hydrogels with non-immunogenic component for cartilage repair. npj Regen. Med. **6**, 54 (2021)
59. Rana, M.M., De la Hoz, S.H.: Tuning the properties of PNIPAm-based hydrogel scaffolds for cartilage tissue engineering. Polym. Basel **13**(18), 3154 (2021)
60. Liu, W., Madry, H., Cucchiarini, M.: Application of alginate hydrogels for next-generation articular cartilage regeneration. J. Mol. Sci. **23**(3), 1147 (2022)
61. Chen, Y., Song, J., Wang, S., Liu, W.: PVA-based hydrogels: promising candidates for articular cartilage repair. Macromol. Biosci. **21**(10), 2100147 (2021)
62. Studenovská, H., Vodička, P., Proks, V., Hlučilová, J., Motlík, J., Rypáček, F.: Synthetic poly(amino acid) hydrogels with incorporated cell-adhesion peptides for tissue engineering. J. Tissue Eng. Regen. Med. **4**(6), 454–463 (2010)
63. Chimisso, V., Garcia, M.A.A., Avsar, S.Y., Dinu, I.A., Palivan, D.G.: Design of bio-conjugated hydrogels for regenerative medicine applications: from polymer scaffold to biomolecule choice. Mol. **25**(18), 4090 (2020)
64. Zhu, M., Zhong, W., Cao, W., Zhang, Q., Wu, G.: Chondroinductive/chondroconductive peptides and their-functionalized biomaterials for cartilage tissue engineering. Bioact. Mater. **9**, 221–238 (2022)
65. Skaalure, S.C., Chu, S., Bryant, S.J.: An enzyme-sensitive PEG hydrogel based on aggrecan catabolism for cartilage tissue engineering. Adv. Healthc. Mater. **4**(3), 420–431 (2015)
66. Gresham, R.C.H., Bahney, C.S., Leach, J.K.: Growth factor delivery using extracellular matrix-mimicking substrates for musculoskeletal tissue engineering and repair. Bioact. Mater. **6**(7), 1945–1956 (2021)
67. Zhang, L., Hu, J., Athanasiou, K.A.: The role of tissue engineering in articular cartilage repair and regeneration. Crit. Rev. Biomed. Eng. **37**(1–2), 1–57 (2009)
68. Fox, A.J.S., Bedi, A., Rodeo, S.A.: The basic science of articular cartilage structure, composition, and function. Sports Health. **1**(6), 461–468 (2009)
69. Ertan, A.B., Yılgor, P., Bayyurt, B., Çalıkoğlu, A.C., Kaspar, C., Kök, F.N., Kose, G.T., Hasirci, V.: Effect of double growth factor release on cartilage tissue engineering. J. Tissue Eng. Regen. Med. **7**(2), 149–160 (2013)
70. Gonzalez-Fernandez, T., Tierney, E.G., Cunniffe, G.M., O'Brien, F.J., Kelly, D.J.: Gene delivery of TGF-β3 and BMP2 in an MSC-laden alginate hydrogel for articular cartilage and endochondral bone tissue engineering. Tissue Eng. Part A. **22**(9–10), 776–787 (2016)
71. Choi, B., Kim, S., Fan, J., Kowalski, T., Petrigliano, F., Evseenko, D., Lee, M.: Covalently conjugated transforming growth factor-β1 in modular chitosan hydrogels for the effective treatment of articular cartilage defects. Biomater. Sci. **3**(5), 742–752 (2015)

# Chapter 11
# Mimicked 3D Scaffolds for Articular Cartilage Surgery

Mimicked 3D scaffolds are performance biomaterials for articular cartilage surgery in both cartilages, without and with included subchondral bone tissue defect (Fig. 11.1). In early, monophasic scaffolds are mimicked for fitting to cartilage tissue. In monophasic scaffolds, a suitable geometry, morphology, and function need to be created for use in the defect site of cartilage [1]. The scaffolds can be used with or without cell encapsulation before implantation [2, 3]. For the latter, biphasic scaffolds are mimicked for matching to cartilage and bone tissue. Normally, biphasic scaffolds have a three-dimensional (3D) porous phase for the bone part and a hydrogel part for the cartilage [4, 5]. Some biphasic scaffolds are created into the 3D porous structural part affixed to the membrane. For these scaffolds, the porous structure is in contact to cartilage tissue. On the other hand, the membrane acts as the barrier to prevent cell invasion, from synovial fluid into the porous structure that leads to disturbance of tissue formation. In this chapter, mimicked 3D scaffolds are described as well as their requirements fitting to this type of surgery. Furthermore, mimicked 3D scaffolds are explained as to their structures and functions, which are similar to cartilage tissue.

## 11.1 3D Scaffolds for Cartilage Defect Without Bone Tissue

3D scaffolds play the important role as structures to promote cell behaviors: adhesion, proliferation, migration, and differentiation in the enhancement of cartilage tissue formation at defect areas. In the case of defects without bone tissue, 3D scaffolds have the performance for assisting in the promotion behaviors of chondrocyte, which leads to enhancement of cartilage tissue formation. Moreover, these 3D scaffolds show performance to maintain the contoured shape at the defect area during new tissue formation. This contributes to regular cartilage formation. These 3D scaffolds are created into two types: 1) monophasic and 2) biphasic form.



J. Meesane, *Mimicked Tissue Engineering Scaffolds for Maxillofacial and Articular Cartilage Surgery*, Engineering Materials,
https://doi.org/10.1007/978-981-19-7830-2_11

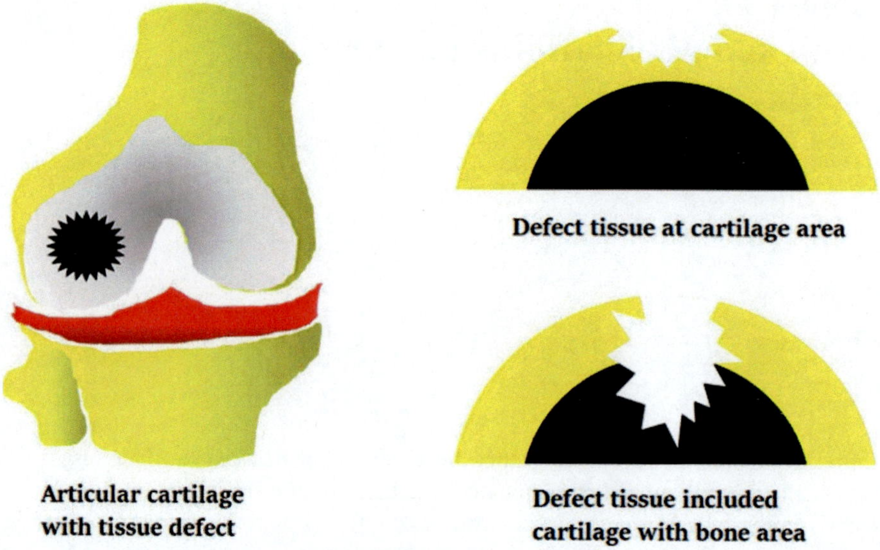


**Fig. 11.1** Articular cartilage defect without and with included bone tissue defect

For monophasic 3D scaffolds, they are created into the porous structural specimen shaped for fitting into the defect space. Before insertion of these scaffolds, the debris at the defect area are removed. Then, the shaped specimen is inserted into the defect area (Fig. 11.2a). This is performed via two methods: (1) scaffold with and (2) without cells (Fig. 11.2b). In the case of scaffolds without cells, they are directly inserted into the defect area. This shows that chondrocytes migrate and biological signals diffuse from the boundary, which is in contact with the cartilage tissue inside the scaffold. In the case of scaffolds with cells, these are seeded with cells: chondrocytes or stem cells, before being inserted into the defect area. In this case, biological signals from the boundary diffuse inside of the scaffolds.

For monophasic scaffolds in cartilage tissue defects, they have a problem regarding cell invasion from synovial fluid to the cartilage defect during tissue formation. Therefore, to solve this problem, the biphasic scaffold of sponge is affixed onto the membrane, and this is created for prevention of cell invasion from synovial fluid (Fig. 11.3a) [6]. Furthermore, this scaffold is designed for incorporation with the method of bone microcracking (Fig. 11.3b). In the use of this method, the debris at the cartilage tissue defect is firstly removed. Then, the removed cartilage defect is made for microcracking, acting as the channel for blood cells and bone marrow stem cell migration from the bone to the cartilage. For the membrane, it shows a barrier function to prevent the diffusion of synovial fluid into the sponge. Furthermore, this membrane demonstrates as the guided cartilage tissue formation.

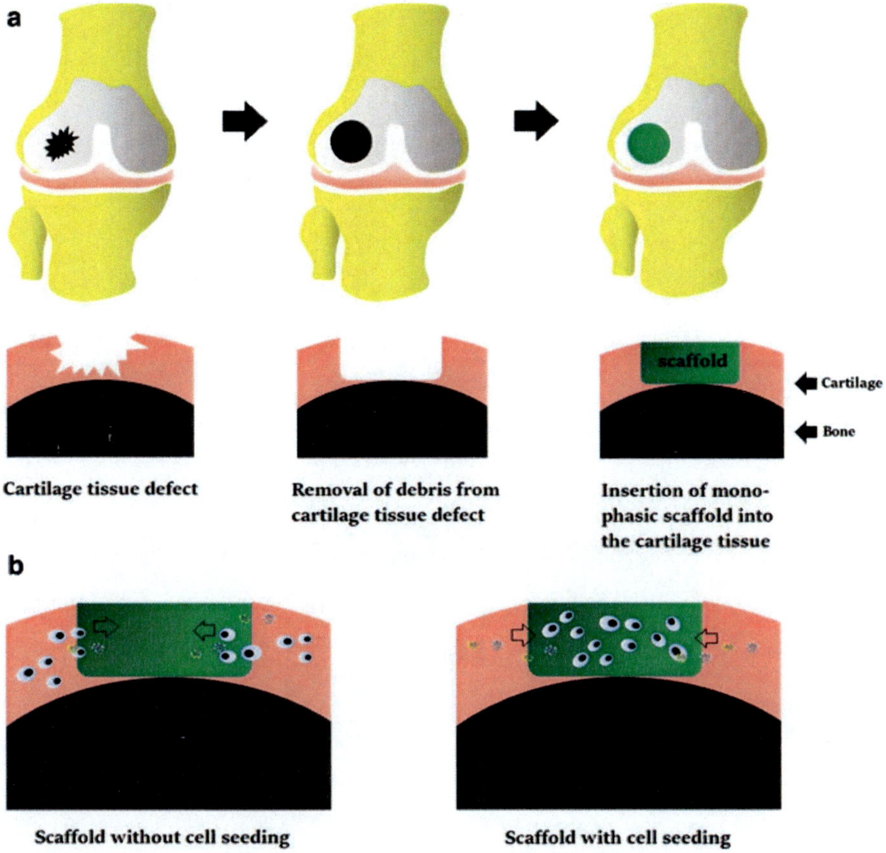


**Fig. 11.2** **a** Monophasic scaffold insertion to the cartilage tissue defect and **b** scaffold without and with cell seeding at cartilage tissue defect

## 11.2 3D Scaffolds for Cartilage with Bone Defects

In the case of cartilage with bone defects, there are two parts of tissue. Cartilage tissue has more softness and elasticity than bone; additionally, it also has more water uptake than bone tissue, whereas bone tissue shows more strength and stiffness than cartilage. For surgical application, some cases use two scaffolds for cartilage and bone defects. Bone scaffolds are firstly placed into the defect. Then, a bone scaffold is covered with a cartilage scaffold. This application has the problem of multistepping during the operation, which is non-convenient for surgery. Hence, biphasic scaffolds were created to solve this problem. Biphasic scaffolds act as the structure for cell behaviors, leading to enhancement of tissue formation of both cartilage and bone. Biphasic scaffolds are specifically designed for maintaining contour shape at the defect space of cartilage and bone. For the operation, at this cartilage defect, the

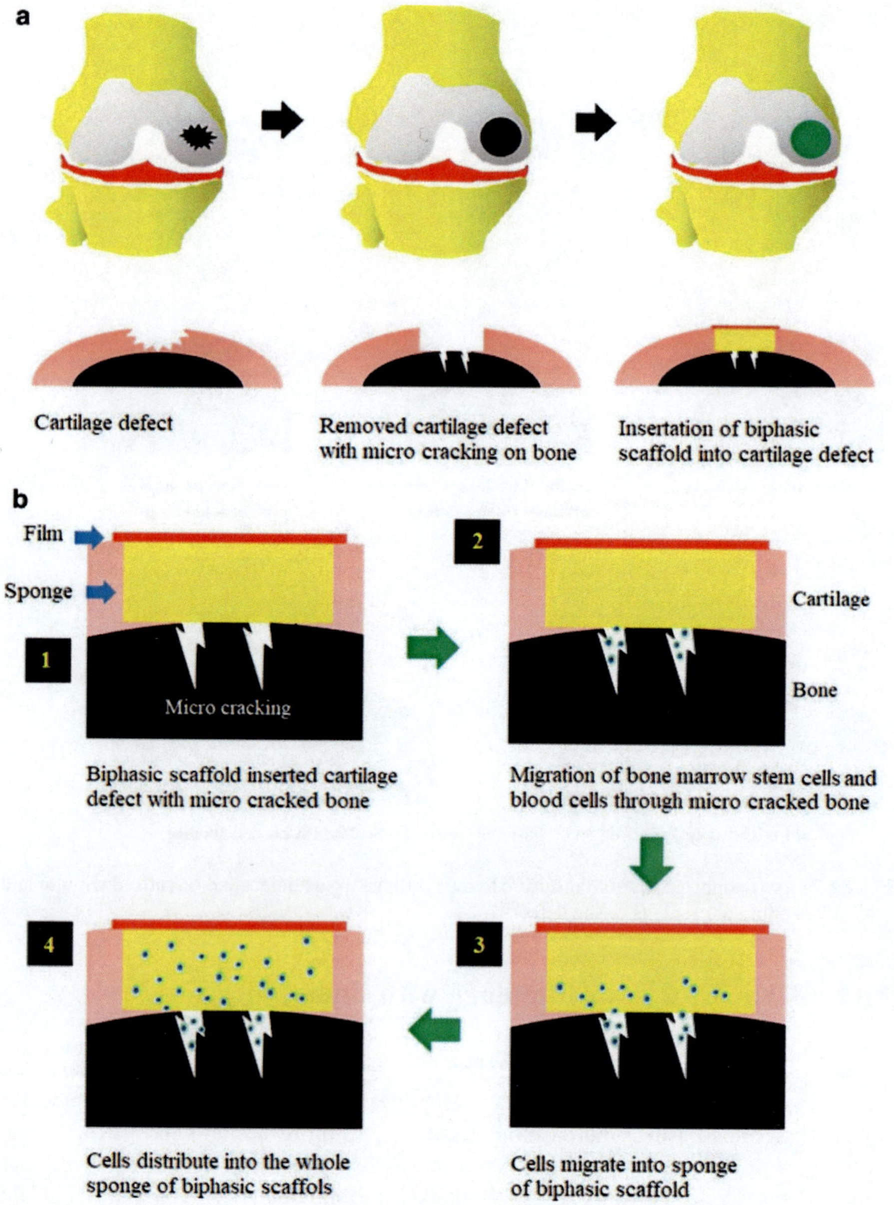


**Fig. 11.3** **a** Insertion of a biphasic scaffold into the cartilage defect, with microcracking of bone and **b** cell behaviors in a biphasic scaffold at the cartilage defect, with microcracking of bone

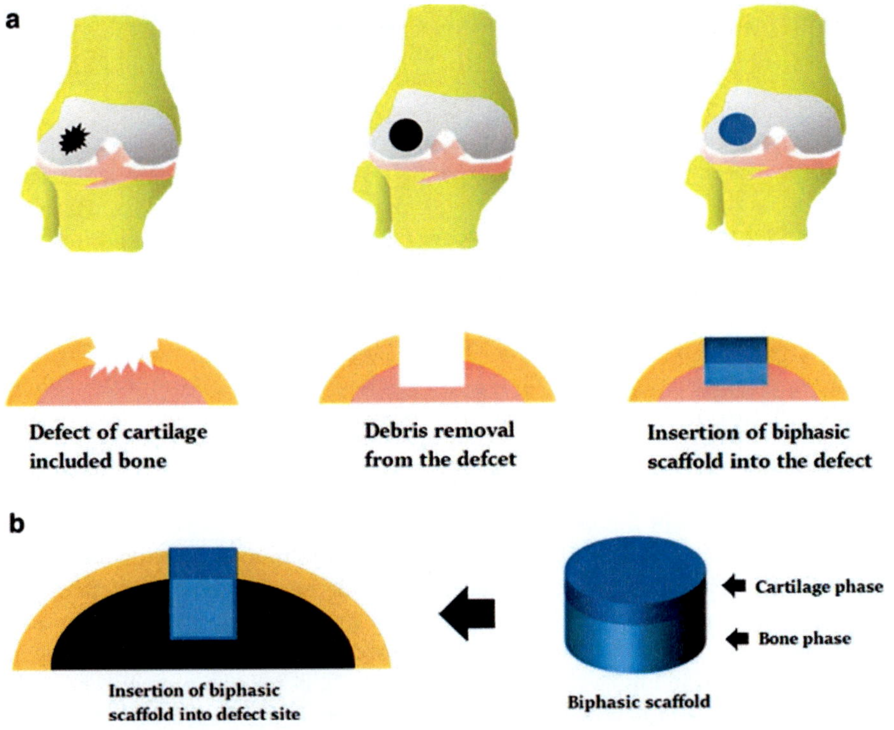


**Fig. 11.4** **a** Articular cartilage surgery with a biphasic scaffold at the defect included with bone and **b** insertion of biphasic scaffold at the defect of articular cartilage

debris at the cartilage and bone site are removed. Then, a biphasic scaffold is inserted into the defect (Fig. 11.4a). Biphasic scaffolds are constructed from two parts: (1) cartilage and (2) bone that show structures and functions similar to natural tissue (Fig. 11.4b).

Normally, biphasic scaffolds need to be in contact with bone marrow mesenchymal stem cells (MSCs) in the bone part and chondrocytes in the cartilage layer. During tissue regeneration, the MSCs can migrate to the interfacial area between the bone and cartilage. At the interfacial area, MSCs interact with the chondrocytes which interfere with each other. This interference causes irregular tissue engineering [7]. Furthermore, chondrocytes can leak from the cartilage layer to the outside, which then come in contact with cells in synovial fluid [84, 85]. The contact between the chondrocytes and cells in synovial fluid leads to irregular tissue engineering [8, 9]. Therefore, the design of the geometry, structure, and functionality is important in the performance of monophasic or biphasic scaffolds.

## 11.3 Mimicked 3D Scaffolds for Articular Cartilage Surgery

For articular cartilage surgery, mono-and biphasic scaffolds are used to promote tissue formation at the defect. To enhance their performance, scaffolds are created so as to be similar to that of natural tissue. Mimicking is the approach to create those scaffolds. Normally, mimicked scaffolds are focused to have structure and function similar to the microenvironment: extracellular matrix and growth factors of natural tissue.

### *11.3.1 Mimicking in Structure of Cartilage Tissue*

For mimicking in structure of cartilage tissue, scaffolds are created similar to the microenvironment, with nano-, micro-, and macro-scale. For the nano-scale or molecular level, polypeptide of collagen type is the model to mimic the structure of the scaffolds [10]. The peptide is synthesized before being directly used to fabricate scaffolds. Some scaffolds are created with blending or grafting of polypeptide, with other natural or synthetic polymers [11]. Furthermore, amino acid motif of collagen type is synthesized to mimic molecular structure [12]. For application, this motif is mixed with either natural or synthetic polymers before being fabricated into a scaffold. In the other application, the motif is conjugated with the polymers. Then, the motif conjugated polymers are fabricated into a scaffold. Some scaffolds are made from mono-or disaccharide of glycosaminoglycan incorporated polymer, which is mimicking the molecular structure [13].

For micro-or morphological structures, a fibril network of collagen is the guidance to mimic scaffold similar to the extracellular matrix. Collagen is prepared into solution before being adjusted into the certain condition, which promotes the fibril network [14]. The mimicked fibril network, which is more similar to the cartilage extracellular matrix, has added glycosaminoglycan particularly hyaluronic acid [15]. This shows morphology of the fibril network in the matrix of glycosaminoglycan. Based on this morphology, some scaffolds are made from other natural or synthetic fibril networks, which are in the polymer matrix [16]. This shows a mimicked morphological structure similar to the extracellular matrix. Other morphology, porous structures, which are similar to the cartilage extracellular matrix, can also be used for mimicked scaffolds [17]. Mainly, these scaffolds demonstrate a pore size which is suitable for chondrocyte [18]. Furthermore, the pores need the inter-connective structure, as it assists in the diffusion of nutrients, growth factors, and cytokines. This contributes to the promotion of new tissue formation.

For macro-structures, monophasic and biphasic scaffolds are constructed, based on the mimicking similar to the anatomy of cartilage [6]. Monophasic scaffolds are mainly created to be similar to cartilage tissue. In the case of biphasic scaffolds, they are designed for fitting to the macro-structure of cartilage, which connects to bone. For this design, the cartilage porous scaffold is bonded with a bone porous scaffold.

To complete the mimicked scaffolds, similar to the extracellular structure, they need to create function similar to cartilage tissue. This is explained in the next section.

### *11.3.2 Mimicking in Function of Cartilage Tissue*

To make complete scaffolds similar to natural tissue, they need the mimicking of function similar to that of cartilage tissue. The mimicked function which is required for mimicking is physical, mechanical, and biological properties and performance. For physical performance, water uptake is important for mimicking of cartilage scaffolds. This is because cartilage has a high water content, which has the role regarding the nutrient diffusion during tissue formation [19]. To mimic this performance, scaffolds have to be fabricated from hydrophilic polymers with their molecular flexibility [20]. Furthermore, these scaffolds must show large pore sizes, with inter-connective structures, this being the cause of high water uptake [21].

In the case of their mechanical performance, the compressive strength of cartilage is mimicked to create the scaffolds. This is because the compressive strength is the main biomechanic condition of cartilage tissue [22]. To mimic the compressive strength, polymers with molecular flexibility are selected to fabricate scaffolds similar to cartilage [20]. Furthermore, to enhance the crystal content of the polymers, in this instance, is the approach to adjust the mechanical strength of the scaffold [23].

For biological performance, mimicked scaffolds are created with the biodegradation rate being similar to the extracellular matrix, which is degraded by enzymes during tissue formation. This is created with biodegradable polymers that match to tissue regeneration rates. For instance, natural polymers of collagen, hyaluronic acid, gelatin, alginate, chitosan, and silk fibroin are often used to mimic performance of biodegradation [24, 25]. These natural polymers also have performance of non-toxicity that is fitting for tissue formation.

Furthermore, biological performance for cell recognition is often mimicked for scaffolds. In this case, components of the extracellular matrix are selected as base materials to fabricate scaffolds. For example, collagen and hyaluronic acid are popular base materials for mimicked scaffolds. Moreover, the cell recognized motif in components of ECMs is selected to mimic biological performance of cell recognition. For instance, cell recognized motif of RGD and YIGSR is often used to conjugate or graft onto/into polymers. These conjugated polymers demonstrate the potential to induce tissue formation.

To completely mimic structure and function, suitable fabrication is required for scaffolds. Hence, fabrication is explained in the next section.

## 11.4 Construction of Mimicked 3D Scaffolds

Before construction of mimicked 3D scaffolds, they need to be fabricated into suitable structures with function via the basic methods: particle leaching, freeze drying, and 3D printing. [This was explained in the chapter 5] After fabrication, 3D scaffolds have to be constructed so as to fit to the defect area at cartilage.

The fitted construction of either monophasic or biphasic mimicked 3D scaffolds can be simply classified into two parts: the cartilage part and bone part. For the cartilage part, scaffolds need to be fabricated into a porous structure that can support cell formation into a spherical structure similar to native cartilage tissue [2, 18, 25]. Furthermore, these scaffolds need to resist the compressive forces during movement and degrade within a specific period of time as the new tissue forms [18, 25, 26]. For the bone part, scaffolds need to have both the physical and mechanical properties to maintain the stability and resist the compressive forces during tissue regeneration [25]. These scaffolds also need to have the biological functionality to induce new bone formation. Furthermore, the scaffolds must degrade within a specific time period during new bone formation [26, 27].

In the case of a biphasic design, scaffolds for the bone and cartilage parts have to be combined. The scaffolds can be connected through physical interaction by diffusion of the bone and cartilage parts [28]. Diffusion leads to the formation of molecular interaction at the interface of the bone and cartilage scaffolds [29]. Therefore, the induction of diffusion needs consideration for biphasic scaffold construction. Similar molecular structures and characteristics of the bone and cartilage parts can induce diffusion that then leads to molecular interaction. Furthermore, the molecular interaction by hydrogen bonding, electrostatic forces, and hydrophilic-hydrophilic and hydrophobic-hydrophobic interactions can enhance the connection between the bone and cartilage parts (Fig. 11.5).

In addition to the physical interaction, physical inter-locking, via the surface roughness, is also considered in the design of biphasic scaffold construction. A high surface roughness can enhance the adhesion between the two parts (Fig. 11.6).

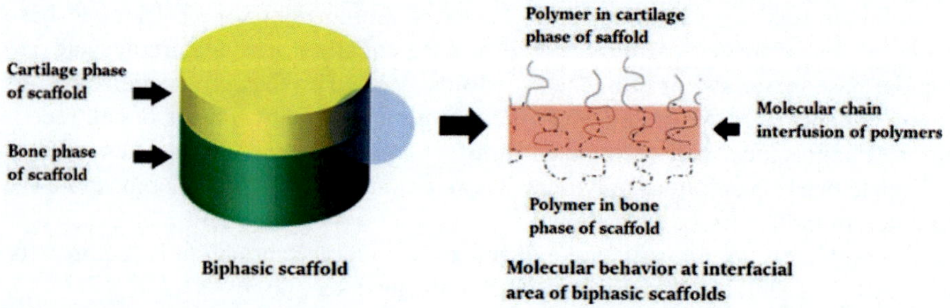


**Fig. 11.5** Molecular interactions at the interface of the bone and cartilage parts of constructed biphasic scaffolds

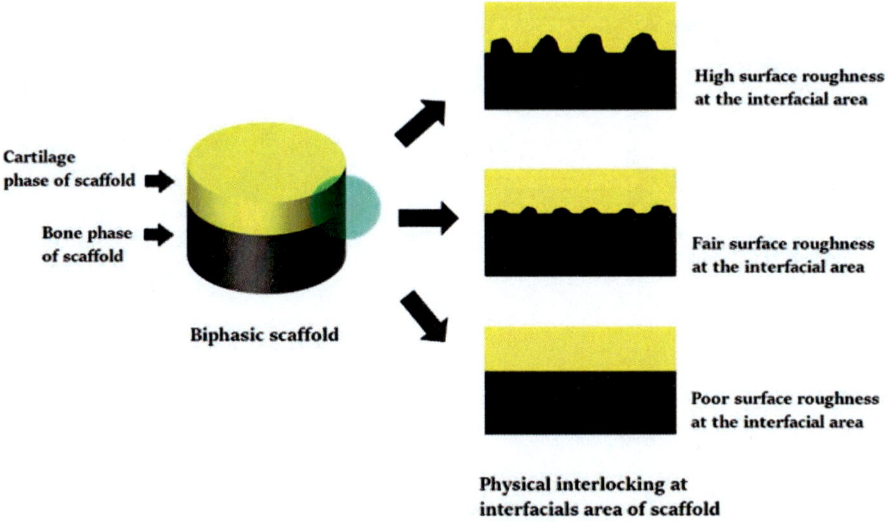


**Fig. 11.6** Physical inter-locking formation, based on the surface roughness at the interfacial phase of a biphasic scaffold construction

## 11.5 Modification in Mimicked 3D Scaffolds

Mimicked 3D scaffolds aim to enhance new tissue formation; hence, they need further modification to promote their potential. There are two classifications: structural modification and functional modification. After fabrication, into either a monophasic or biphasic geometry, the morphological structure of scaffolds can be modified with other materials, particularly by polymers, into other geometries [30]. Generally, the geometry of scaffolds exhibits a porous structure, and these porous structures have interconnections that can induce new tissue formation [31]. Modified geometries of porous structures are simply created by immersion in polymer solutions, with different concentrations, before freeze drying [32]. The structure of freeze-dried scaffolds appears as a network of main pores in their structures (Fig. 11.7).

A structure, with a network of pores from the modification of freeze drying, enhances cell adhesion and spreading during new tissue formation. Furthermore, a structure with a network of pores can induce physical stability that can maintain a specific contour shape at the defect site during new tissue formation. Different network porous structures have different physical and mechanical properties [33]. Therefore, construction of specific network porous structures, which fit the bone or cartilage defect, is important in the design of performance scaffolds.

Besides modification of porous scaffolds by immersion in a polymer solution, some scaffolds can be modified with chemical reagents. For instance, silk fibroin porous scaffolds can be treated with different concentrations of alcohol solutions [34]. Alcohol can induce structural transformation from random coils and alpha-helix to β-sheet [35]. For different concentrations of alcohol, there are different potential

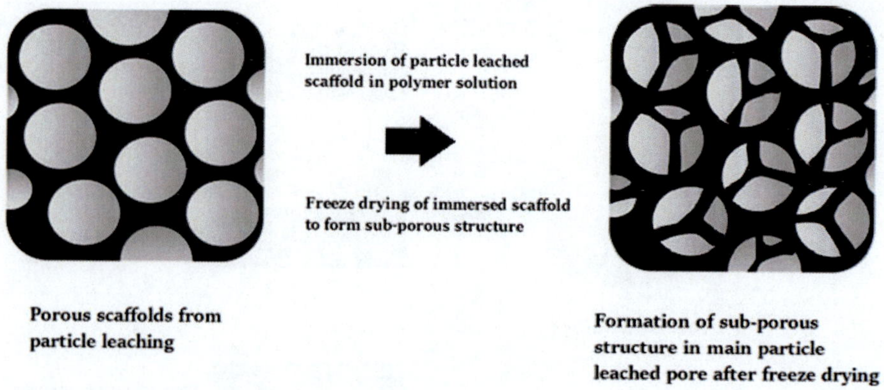


**Fig. 11.7** Modified geometry of porous scaffolds

transformations that lead to different organizations of the outermost surface [36, 37]. A different outermost formation leads to a different roughness characteristic. In the case of high roughness, cells can adhere firmly to the surface. Furthermore, a high surface roughness can induce cell migration that leads to the enhancement of new tissue formation.

Functional modifications of bone or cartilage part scaffolds can be done by adding other polymers that can induce a physical, mechanical, or biological functionality [38, 39]. The physical and mechanical functionalities of scaffolds can be modified with the addition of polymers that enhance the strength, stiffness, toughness, swelling behavior, and water uptake.

For instance, calcium phosphate is a popular mineral that is often added to scaffolds to induce physical and mechanical functionality [40]. Generally, the scaffolds with calcium phosphate have high stiffness, strength, and physical stability that are suitable for the bone part. Furthermore, calcium phosphate also has a biological functionality to induce new bone formation.

In the case of physical and mechanical modification of the cartilage part, polymers have often been added to scaffolds that have chemical groups of hydrophilicity and high molecular weight [41]. The chemical groups can help to induce water uptake and high swelling behavior similar to native cartilage. The molecular weight can enhance the resistance to compressive forces [42]. It is popular to add HA in scaffolds, because HA has carboxylic groups that have the unique functionality to reserve a high amount of water [106]. Furthermore, based on its high molecular weight HA can also resist high compressive forces similar to that of native cartilage [43].

For the case of biological modification of the cartilage part, some polymers can be added to scaffolds that induce tissue regeneration [44]. Some polymers can be added into the scaffolds to design biodegradation matching that of new tissue formation [45, 46]. For instance, positive charged chitosan can be added to scaffolds to induce cell adhesion and biodegradation [47, 48]. Furthermore, the incorporation of biological signals in mimicked 3D scaffolds is a method to enhance biological functionality, cell adhesion, proliferation, migration, and differentiation [49, 50]. Those biological functionalities lead to inducement of tissue regeneration. The two main approaches to incorporate biological signals in biodegradable polymers are encapsulation and non-encapsulation [51]. Encapsulated biological signals can be added into a polymer solution before fabrication into scaffolds. In this method, the signals are distributed throughout the whole texture of the polymer scaffold. Furthermore, at the surface of the scaffolds, the signals are covered with polymers, and these signals will be active when those polymers degrade. Then, the signals emerge and are active with the cells during new tissue formation [52]. However, the problem with this method is the dysfunction of the signals during scaffold fabrication. Therefore, to solve this problem the signals are immobilized on the surface of scaffolds [53]. This method can reduce the dysfunction of the signals. Nevertheless, the problem with the immobilization method is the burst release of signals, which are active in the early period of new tissue formation. Therefore, it is important to create a sustainable release of the signals to solve said problem and to enhance the performance of scaffolds [54].

Encapsulation of the biological signals is a method that can both maintain the functionality and sustain the release of the signals [55]. The signals are encapsulated with biodegradable polymers. Afterward, the encapsulated signals are mixed in a polymer solution before fabrication into the scaffolds. The encapsulated signals in the scaffolds are distributed throughout the whole texture of the scaffolds. This enables the scaffolds to perform sustainable release of the signals, and the signals can function within a specific period of time after degradation of the polymers during new tissue formation. Another method is to have the scaffolds immobilized with encapsulated signals. The encapsulated signals can release rapidly from the surface of the scaffolds. Then, these signals can function for a specific period after degradation of the polymers.

Based on these methods, a mimicked profile release of the biological signals can be created with different protocols of scaffold fabrication with encapsulation. For instance, in the early period the function of encapsulated signals can be immobilized on the surface of the scaffolds. On the other hand, a prolonged functioning period of encapsulated signals can be mixed in the polymer solution before fabrication into scaffolds. Examples of scaffolds incorporated with biological signals are shown in Fig. 11.8.

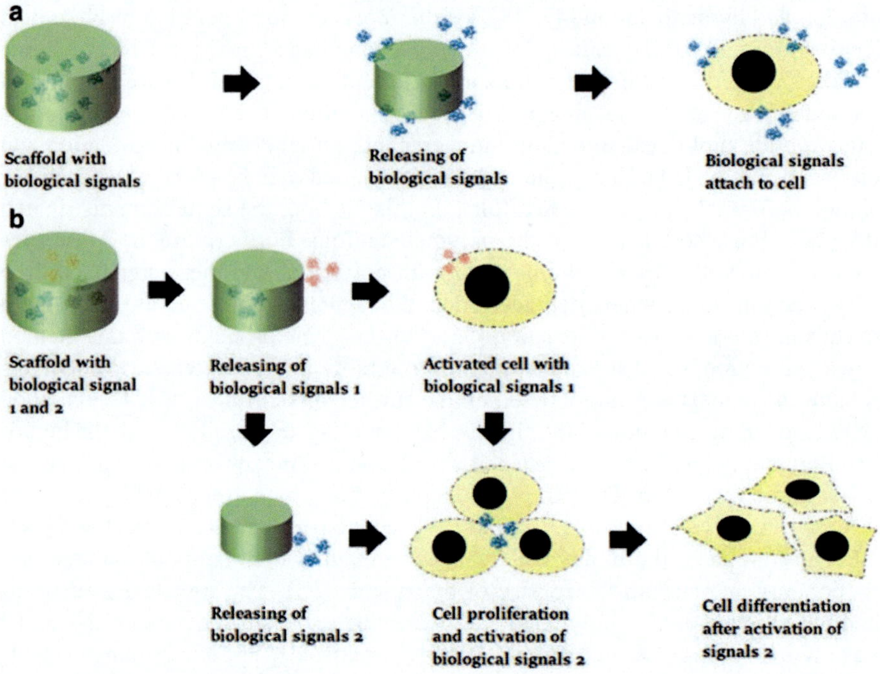


**Fig. 11.8** Examples of scaffolds incorporated with biological signals: Scaffold incorporated with early biological signals **(a)** and programming of biological signals for cell regulation into tissue regeneration **(b)**

## References

1. Izadifar, Z., Chen, X., Kulyk, W.: Strategic design and fabrication of engineered scaffolds for articular cartilage repair. J. Funct. Biomater. **3**(4), 799–838 (2012)
2. Jaipaew, J., Wangkulangkul, P., Meesane, J., Raungrut, P., Puttawibul, P.: Mimicked cartilage scaffolds of silk fibroin/hyaluronic acid with stem cells for osteoarthritis surgery: morphological, mechanical, and physical clues. Mater. Sci. Eng. C Mater. **64**, 173–182 (2016)
3. Kim, I.G., Ko, J., Lee, H.R., Do, S.H., Park, K.: Mesenchymal cells condensation-inducible mesh scaffolds for cartilage tissue engineering. Biomaterials **85**, 18–29 (2016)
4. Li, X., Ding, J., Wang, J., Zhuang, X., Chen, X.: Biomimetic biphasic scaffolds for osteochondral defect repair. Regen. Biomater. **2**(3), 221–228 (2015)
5. Kazunori, S., Yu, M., Christopher, D.M., Hideki, Y., Norimasa, N.: Osteochondral tissue engineering with biphasic scaffold: current strategies and techniques. Tissue Eng. Part B Rev. **20**(5), 468–476 (2014)
6. Panjapheree, K., Kamonmattayakul, S., Meesane, J.: Biphasic scaffolds of silk fibroin film affixed to silk fibroin/chitosan sponge based on surgical design for cartilage defect in osteoarthritis. Mater. Design **141**, 323–332 (2018)
7. Vinatier, C., Bouffi, C., Merceron, C., Gordeladze, J., Brondello, J.M., Jorgensen, C., Weiss, P., Guicheux, J., Noël, D.: Cartilage tissue engineering: towards a biomaterial-assisted mesenchymal stem cell therapy. Curr. Stem Cell Res. Ther. **4**(4), 318–329 (2009)
8. Bonnet, C.S., Walsh, D.A.: Osteoarthritis, angiogenesis and inflammation. Rheumatology **44**(1), 7–16 (2005)

9. Rainbow, R., Ren, W., Zeng, L.: Inflammation and joint tissue interactions in oa: implications for potential therapeutic approaches. Arthritis **2012**, 741582 (2012)
10. Parmar, P.A., Chow, L.W., St-Pierre, J.P., Horejs, C.M., Peng, Y.Y., Werkmeister, J.A., Ramshaw, J.A.M., Stevens, M.M.: Collagen-mimetic peptide-modifiable hydrogels for articular cartilage regeneration. Biomaterials **54**, 213–225 (2015)
11. Klimek, K., Ginalska, G.: Proteins and peptides as important modifiers of the polymer scaffolds for tissue engineering applications—a review. Polym. Basel **12**, 844 (2020)
12. Yi, J., Liu, Q., Zhang, Q., Chew, T.G., Ouyang, H.: Modular protein engineering-based biomaterials for skeletal tissue engineering. Biomaterials **282**, 121414 (2022)
13. Sánchez-Téllez, D.A., Téllez-Jurado, L., Rodríguez-Lorenzo, L.M.: Hydrogels for cartilage regeneration, from polysaccharides to hybrids. Polym. Basel **9**(12), 671 (2017)
14. Kyle, S., Aggeli, A., Ingham, E., McPherson, M.J.: Production of self-assembling biomaterials for tissue engineering. Trends Biotechnol. **27**(7), 423–433 (2009)
15. Romanelli, S.M., Knoll, G.A., Santora, A.M., Brown, A.M., Banerjee, I.A.: Comparison of engineered peptide-glycosaminoglycan microfibrous hybrid scaffolds for potential applications in cartilage tissue regeneration. Fibers **3**(3), 265–295 (2015)
16. Irawan, V., Sung, T.C., Higuchi, A., Ikoma, T.: Collagen scaffolds in cartilage tissue engineering and relevant approaches for future development. Tissue Eng. Regen. Med. **15**(6), 673–697 (2018)
17. Wang, J., Wang, Y., Sun, X., Liu, D., Huang, C., Wu, J., Yang, C., Zhang, Q.: Biomimetic cartilage scaffold with orientated porous structure of two factors for cartilage repair of knee osteoarthritis. Artif. Cell Nanomed. B **47**(1), 1710–1721 (2019)
18. Nava, M.M., Draghi, L., Giordano, C., Pietrabissa, R.: The effect of scaffold pore size in cartilage tissue engineering. J. Appl. Biomater. Func. **14**(3), e223–e229 (2016)
19. Ngadimin, K.D., Stokes, A., Gentile, P., Ferreira, A.M.: Biomimetic hydrogels designed for cartilage tissue engineering. Biomater. Sci. **9**, 4246–4259 (2021)
20. Schüller-Ravoo, S., Teixeira, S.M., Feijen, J., Grijpma, D.W., Poot, A.A.: Flexible and elastic scaffolds for cartilage tissue engineering prepared by stereolithography using poly(trimethylene carbonate)-based resins. Macromol. Biosci. **13**(12), 1711–1719 (2013)
21. Xie, Y., Lee, K., Wang, X., Yoshitomi, T., Kawazoe, N., Yang, Y., Chen, G.: Interconnected collagen porous scaffolds prepared with sacrificial PLGA sponge templates for cartilage tissue engineering. J Mater. Chem. B **9**, 8491–8500 (2021)
22. Koh, Y.G., Lee, J.A., Kim, Y.S., Lee, H.Y., Kim, H.J., Kang, K.T.: Optimal mechanical properties of a scaffold for cartilage regeneration using finite element analysis. J. Tissue Eng. **10**, 2041731419832133 (2019)
23. Chen, Y., Song, J., Wang, S., Liu, W.: PVA-Based Hydrogels: Promising Candidates for Articular Cartilage Repair. Macromol. Biosci. **21**, 2100147 (2021)
24. Zhao, W., Jin, X., Cong, Y., Liu, Y., Fu, J.: Degradable natural polymer hydrogels for articular cartilage tissue engineering. J. Chem. Technol. Biotechnol. **88**(3), 327–339 (2013)
25. Cheng, G., Davoudi, Z., Xing, X., Yu, X., Cheng, X., Li, Z., Deng, H., Wang, Q.: Advanced silk fibroin biomaterials for cartilage regeneration. ACS Biomater. Sci. Eng. **4**(8), 2704–2715 (2018)
26. O'Brien, F.J.: Biomaterials & scaffolds for tissue engineering. Mater. Today **14**, 88–95 (2011)
27. Cheng, N.C., Estes, B.T., Young, T.H., Guilak, F.: Engineered cartilage using primary chondrocytes cultured in a porous cartilage-derived matrix. Regen. Med. **6**(1), 81–93 (2011)
28. Guarino, V., Causa, F., Ambrosio, L.: Porosity and mechanical properties relationship in PCL porous scaffolds. J. Appl. Biomater. Biomech. **5**(3), 149–157 (2007)
29. Walker, J.M., Bodamer, E., Krebs, O., Luo, Y., Kleinfehn, A., Becker, M.L., Dean, D.: Effect of chemical and physical properties on the in vitro degradation of 3d printed high resolution poly(propylene fumarate) scaffolds. Biomacromol **18**(4), 1419–1425 (2017)
30. Li, J.J., Kim, K., Roohani-Esfahani, S.I., Guo, J., Kaplan, D.L., Zreiqat, H.: A biphasic scaffold based on silk and bioactive ceramic with stratified properties for osteochondral tissue regeneration. J. Mater. Chem. B **3**, 5361–5376 (2015)

31. Bharati, A., Wübbenhorst, M., Moldenaers, P., Cardinaels, R.: Effect of compatibilization on interfacial polarization and intrinsic length scales in biphasic polymer blends of PαMSAN and PMMA: a combined experimental and modeling dielectric study. Macromol. **49**(4), 1464–1478 (2016)
32. Kim, J.H., Linh, N.T., Min, Y.K., Lee, B.T.: Surface modification of porous polycaprolactone/biphasic calcium phosphate scaffolds for bone regeneration in rat calvaria defect. J. Biomater. Appl. **29**(4), 624–635 (2014)
33. Murphy, C.M., O'Brien, F.J.: Understanding the effect of mean pore size on cell activity in collagen-glycosaminoglycan scaffolds. Cell Adh. Migr. **4**(3), 377–381 (2010)
34. Wangkulangkul, P., Jaipaew, J., Puttawibul, P., Meesane, J.: Constructed silk fibroin scaffolds to mimic adipose tissue as engineered implantation materials in post-subcutaneous tumor removal. Mater. Design **106**, 428–435 (2016)
35. Ikeda, R., Fujioka, H., Nagura, I., Kokubu, T., Toyokawa, N., Inui, A., Makino, T., Kaneko, H., Doita, M., Kurosaka, M.: The effect of porosity and mechanical property of a synthetic polymer scaffold on repair of osteochondral defects. Int. Orthop. **33**(3), 821–828 (2009)
36. Nazarov, R., Jin, H.J., Kaplan, D.L.: Porous 3-D scaffolds from regenerated silk fibroin. Biomacromol. **5**(3), 718–726 (2004)
37. Nam, J., Park, Y.H.: Morphology of regenerated silk fibroin: Effects of freezing temperature, alcohol addition, and molecular weight. J. Appl. Polym. Sci. **81**, 3008–3021 (2001)
38. Servoli, E., Maniglio, D., Motta, A., Predazzer, R., Migliaresi, C.: Surface properties of silk fibroin films and their interaction with fibroblasts. Macromol. Biosci. **5**(12), 1175–1183 (2005)
39. Terada, D., Yokoyama, Y., Hattori, S., Kobayashi, H., Tamada, Y.: The outermost surface properties of silk fibroin films reflect ethanol-treatment conditions used in biomaterial preparation. Mater. Sci. Eng. C Mater **58**, 119–126 (2016)
40. Iqbal, M., Xu, X., Li, L.: A review on biodegradable polymeric materials for bone tissue engineering applications. J. Mater. Sci. **44**, 5713–5724 (2009)
41. Sionkowska, A.: Current research on the blends of natural and synthetic polymers as new biomaterials: review. Prog. Polym. Sci. **36**, 1254–1276 (2011)
42. Zhang, X., Chen, Y., Akram, M.Y., Nie, J., Zhu, X.: Preparation of polymer/calcium phosphate porous composite as bone tissue scaffolds. Mater. Sci. Eng. C Mater. Biol. Appl. **70**(Pt 2), 1125–1131 (2017)
43. Zhu, J., Marchant, R.E.: Design properties of hydrogel tissue-engineering scaffolds. Expert Rev. Med. Devices **8**(5), 607–626 (2011)
44. White, L.J., Hutter, V., Tai, H., Howdle, S.M., Shakesheff, K.M.: The effect of processing variables on morphological and mechanical properties of supercritical CO2 foamed scaffolds for tissue engineering. Acta Biomater. **8**, 61–71 (2012)
45. Levett, P.A., Hutmacher, D.W., Malda, J., Klein, T.J.: Hyaluronic acid enhances the mechanical properties of tissue-engineered cartilage constructs. PLoS ONE **9**(12), e113216 (2014)
46. Guo, B., Lei, B., Li, P., Ma, P.X.: Functionalized scaffolds to enhance tissue regeneration. Regen. Biomater. **2**(1), 47–57 (2015)
47. Sant, S., Iyer, D., Gaharwar, A., Patel, A., Khademhosseini, A.: Effect of biodegradation and de novo matrix synthesis on the mechanical properties of VIC-seeded PGS-PCL scaffolds. Acta Biomater. **9**(4), 5963–5973 (2013)
48. Hongbo, Z., Li, Z., Wenjun, Z.: Control of scaffold degradation in tissue engineering: a review. Tissue Eng. Part B Rev. **20**(5), 492–502 (2014)
49. Croisier, F., Jérôme, C.: Chitosan-based biomaterials for tissue engineering. Eur. Polymer J. **49**, 780–792 (2013)
50. Rodríguez-Vázquez, M., Vega-Ruiz, B., Ramos-Zúñiga, R., Saldaña-Koppel, D.A., Quiñones-Olvera, L.F.: Chitosan and its potential use as a scaffold for tissue engineering in regenerative medicine. Biomed. Res. Int. **2015**, 821279 (2015)
51. Dang, P.N., Dwivedi, N., Phillips, L.M., Yu, X., Herberg, S., Bowerman, C., Solorio, L.D., Murphy, W.L., Alsberg, E.: Controlled dual growth factor delivery from microparticles incorporated within human bone marrow-derived mesenchymal stem cell aggregates for enhanced bone tissue engineering via endochondral ossification. Stem Cells Trans. Med. **5**, 206–217 (2016)

52. Masaeli, E., Karamali, F., Loghmani, S., Eslaminejad, M.B., Nasr-Esfahani, M.H.: Bioengineered electrospun nanofibrous membranes using cartilage extracellular matrix particles. J. Mater. Chem. B **5**, 765–776 (2017)
53. Lee, K., Silva, E.A., Mooney, D.J.: Growth factor delivery-based tissue engineering: general approaches and a review of recent developments. J. R. Soc Interface **8**(55), 153–170 (2011)
54. Ma, Z., Gao, C., Gong, Y., Shen, J.: Cartilage tissue engineering PLLA scaffold with surface immobilized collagen and basic fibroblast growth factor. Biomaterials **26**(11), 1253–1259 (2005)
55. Sohier, J., Moroni, L., van Blitterswijk, C., de Groot, K., Bezemer, J.M.: Critical factors in the design of growth factor releasing scaffolds for cartilage tissue engineering. Expert Opin. Drug Deliv. **5**(5), 543–566 (2008)

# Chapter 12
# Mimicked 2D Scaffolds in Articular Cartilage Surgery

For articular cartilage surgery, hydrogel and 3D scaffolds are the main materials that play the role as structures, with suitable function to enhance tissue formation, particularly in deep defects, with included bone. On the other hand, in partial defects with non-included bone, 2D scaffolds are used to assist for guidance to completed tissue formation. Mimicking is used to create 2D scaffolds, both with similar structures and functions as the microenvironment: extracellular matrix and growth factors of cartilage tissue. In this chapter, the approach of surgery with 2D scaffolds is demonstrated, and mimicked 2D scaffolds in cartilage surgery are explained. Furthermore, fabrication and modification of mimicked 2D scaffolds are described.

## 12.1 2D Scaffolds for Supporting Articular Cartilage Surgery

In the case of 2D scaffolds or membranes, they are often used to articular cartilage surgery for partial tissue defects, with non-included bone [1–5] (Fig. 12.1a). For this surgery, chondrocytes from the patient are initially cultured and expanded in the laboratory, before injected into the defect site. At the defect site, the debris of the defect are removed, before being covered and sealed by a collagen membrane, which acts as a 2D scaffold. After sealing, the chondrocytes from the laboratory are injected beneath the area of the collagen membrane seal (Fig. 12.1b). For this surgery, the membrane shows its function in the prevention of leakage of cells into synovial fluid, as any leakage would lead to irregular tissue formation. Furthermore, this membrane guides and regulates the cells into the cartilage tissue [6].

Another approach for surgery of partial defects is for the surgeons to remove debris during the soft tissue phase, which is in contact with the zone of the bone tissue. Then, the surgeon makes micropores, which are able to channel the migration of bone marrow MSCs to the area where the cartilage tissue has been removed. After



J. Meesane, *Mimicked Tissue Engineering Scaffolds for Maxillofacial and Articular Cartilage Surgery*, Engineering Materials,
https://doi.org/10.1007/978-981-19-7830-2_12

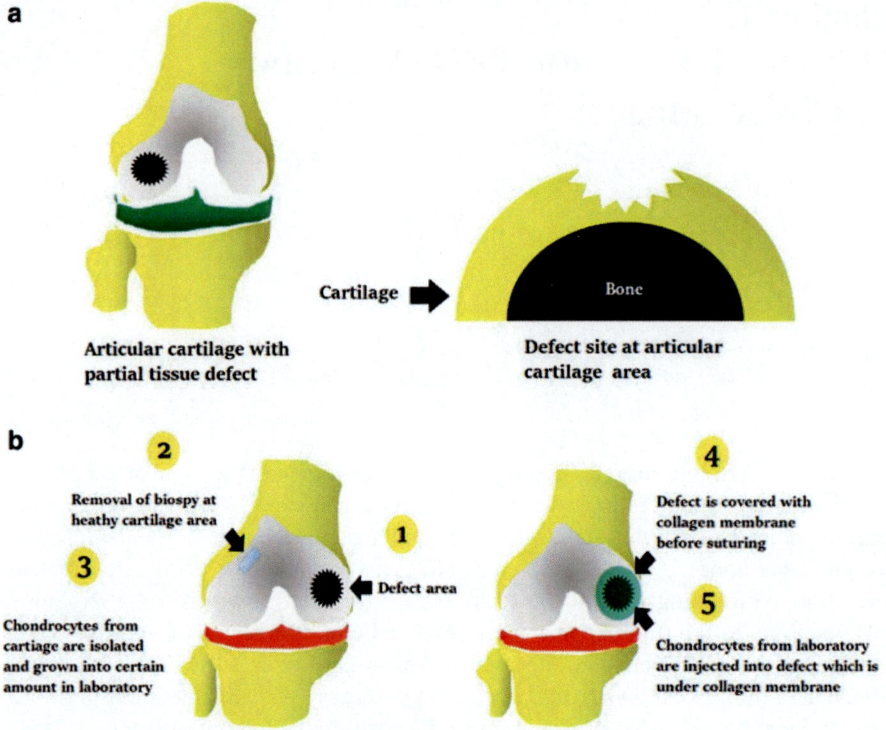


**Fig. 12.1** **a** Partial tissue defect, with non-included bone of articular cartilage and **b** surgery with supporting of the membrane as a 2D scaffold

this, fibrin glue is then coated onto the bone part that is covered by a membrane as illustrated in Fig. 12.2. The membranes serve as a barrier to prevent leaking of the MSCs to an area outside of the cartilage tissue. After MSCs migrate to the cartilage area, that was removed, they can regulate themselves into chondrocytes.

## 12.2 Mimicked 2D Scaffolds for Articular Cartilage Surgery

For the mimicking, 2D scaffolds are created with the structure and function similar to cartilage tissue [7]. As hydrogel and 3D scaffolds, 2D scaffolds are also created to mimic structure of cartilage. First, the nano- or molecular structure of ECMs coupled with growth factors is mimicked for 2D scaffolds. As in the previous chapters, nano-structures are mimicked with the synthesis for partial segments or whole molecules. Collagen type II, hyaluronic acid, bone morphogenetic proteins (BMPs), and transforming growth factor-β (TGF-β) are the molecules often used for inducing

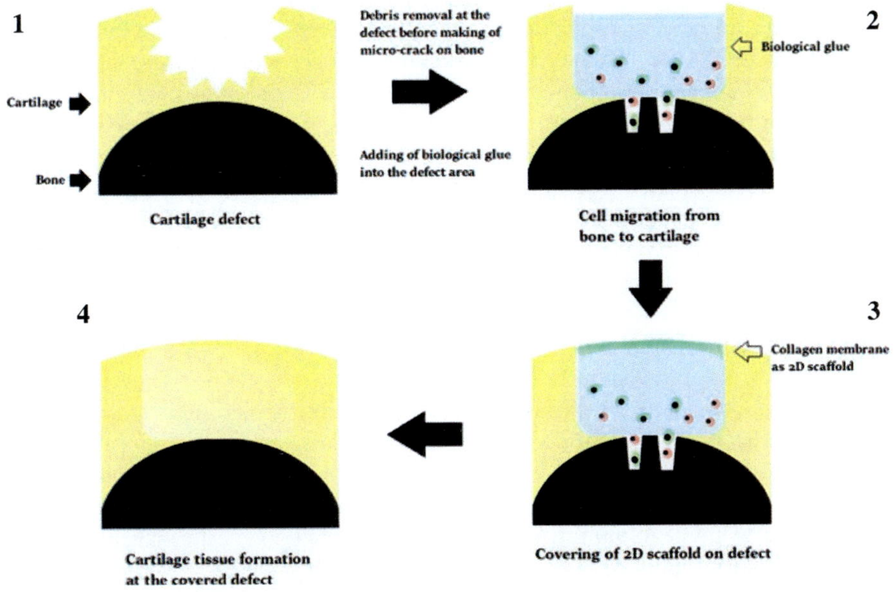


**Fig. 12.2** Surgery of partials defect, with supporting of 2D scaffolds and micropores

of cartilage tissue formation. RGD and YIGSR are the partial segments of ECMs, and growth factors have the role to induce tissue formation. Normally, to mimic the whole molecules or partial segments, protein engineering is used for synthesis. Second, micro- or the morphological structures of ECMs are often mimicked for 2D scaffolds. Collagen type II is the main protein in cartilage tissue, as collagen type II shows network structures of fibrils with different sizes. This is used as a guide for mimicking of fibril structures in ECMs, for 2D scaffolds. Third, macro- or geographical structures at the superficial zone of cartilage are used to guide mimicked 2D scaffolds.

For mimicking of the function, 2D scaffolds are created similar to cartilage tissue. Physical, mechanical, and biological function of cartilage tissue is mimicked for 2D scaffolds. In the case of physical function, wettability is an important performance, and this refers to the potential of 2D scaffolds to assist tissue formation [8]. Wettability of 2D scaffolds has the important role for promoting cell attachment, absorption of nutrients, and biological signals during tissue formation. This leads to enhancement of tissue formation. Furthermore, in the approach of surgery, with 2D scaffold, hydrophilic fibrin glue is filled into the defect. Then, the filled defect is covered with a 2D scaffold. Hence, this 2D scaffold shows a hydrophilic performance fitting to the adhesion of fibrin glue. For mechanical function, because 2D scaffold in articular cartilage surgery is used to cover on the fibrin glue or concentrated chondrocytes, it is placed on the superficial zone. Hence, the 2D scaffold has to mimic the mechanical performance which is similar to the tissue within this zone. This zone is the soft tissue which has the elasticity and toughness to resist the biomechanic of knee joint movement. This is used to be guidance for mimicking of mechanical performance on 2D scaffolds in articular cartilage surgery. Finally, in the case of biological function, 2D scaffolds need to mimic articular cartilage tissue, especially, the microenvironment, which is a biological signal to promote tissue formation. In regard to the microenvironment, ECMs and growth factors are the main components that are often used to mimic said biological function. For instance, collagen type II, BMPs, and TGF-β are mixed with other polymers, so as to create mimicked 2D scaffolds. This is used to be guidance for the mimicking of biological function in the promotion of tissue formation. Furthermore, biodegradation with enzymes on the ECM, during tissue formation, is another mimicking of biological function. These degraded ECMs have the important role in the support of cell proliferation, migration, and proliferation during tissue formation. To mimic this biodegradation, amino acid motif, which is digested with enzymes of the matrix metalloproteinases (MMPs), is conjugated to molecules of the polymers. Then, these conjugated polymers are fabricated into 2D scaffolds, so as to match the degradation matching to the new tissue formation. Nevertheless, to create mimicked 2D scaffolds, with the good performance, suitable fabrication has to be undertaken.

## 12.3 Fabrication in Mimicked 2D Scaffolds

Fabrication for mimicked 2D scaffolds has the important role of maintaining the consistency of their performance. Generally, solution casting and electro-spinning are often used to create 2D scaffolds [9]. For the first, solution casting is a simple method for fabrication of 2D scaffolds. After suitable polymers are selected, they are dissolved into solution. Then, this polymer solution is casted into the mold, before solvent evaporation. For this method, the usual problem is its dense texture, with stiffness, which can be difficult to handle [10]. This is especially true for stiffness, chain polymers that often show dense texture, with high rigidity, which can lead to breaking during handling [11]. This is because of the relaxation of the polymer molecules, which lead to high regular or crystalline formation, thus leading to their dense texture [12]. To solve this problem, plasticizers are added into the polymer solution before being fabricated into 2D scaffolds; this then shows performance of mechanical flexibility [13] and, thus, suitable to be handled. Nevertheless, for some cases the plasticizers are bled during [14], leading to the re-stiffness on 2D scaffolds. To solve this problem, high molecular weight plasticizer is selected, so as to avoid its rapid diffusion that leads to said bleeding [15]. However, solution casting still has the problem on morphology, with low porous structures which are non-sufficient for 2D scaffolds. Elastic and tough polymers are selected as base materials to mimic mechanical function, which is similar to soft tissue [16]. The polymers which have biodegradable performance are often used to mimic biological function [17].

For the second method, electro-spinning is the performance method to fabricate 2D scaffolds. Principally, polymer solutions with charges are spun from the needle and form into fibers with different sizes. Then, the electro-spun fibers are collected into the membrane, which shows suitable structure and performance for a 2D scaffold. Based on electro-spinning, scaffolds are fabricated into fibers similar to the ECM, with this fabrication fitting to mimic the structure of the ECM. Mimicked 2D scaffolds, based electro-spun fibers, show elasticity similar to soft tissue. Furthermore, some mimicked 2D scaffolds are fabricated with biodegradable polymers, so as to mimic biological function.

However, for high performance of scaffolds, it is difficult to mimic scaffolds which are similar to natural tissue; hence, they need modifications to enhance their performance.

## 12.4 Modifications in Mimicked 2D Scaffolds

According to limitations of fabrication, some scaffolds have to be modified for structure or function. For instance, electro-spinning has a limitation for incorporation of biological signals in its fibers during fabrication. This is because these signals are damaged by the high energy generated by the needle during spinning. Hence, to incorporate the biological signals this is managed by modifying the scaffolds after fabrication. Generally, structure and function of 2D scaffolds are modified with either physical or chemical techniques.

For structural modification, mimicked 2D scaffolds have modified surface roughness, which can induce cell regulation [18, 19]. For the physical techniques of modification, some scaffolds have modified surface roughness, via either plasma treatment or other low energy treatments [20]. Different surface roughness depends on the energy of the plasma. Furthermore, the chemical structures on the surface are changed during said treatment. This is because of the gas under condition of treatment, and these changed chemical structures come from the reaction during treatment. For instance, in the case of treatment under $O_2$ the reaction starts with a chemically active species that can be generated from the treatments, particularly in free radicals [21]. Then, the species react with the gas during the treatment (Fig. 12.3). Normally, the chemical groups from this treatment often show hydrophilic functionality, which can promote protein adsorption and cell adhesion [22]. The promoting of proteins and cell adhesion leads to inducement of tissue regeneration.

Based on the generation of chemically active species during the treatments, some molecules are added during treatment, which can operate in the inert gas. Due to the treatment, the molecules are immobilized on the surface of the scaffolds through a reaction with the chemically active species. These immobilized molecules are called "spacers" [23] (Fig. 12.4).

The size, structure, and characteristics of spacers have an effect on protein adsorption, which in turn leads to inducing of cell adhesion during tissue regeneration. For instance, long chains have an effect on the reduction of protein adsorption, whereas short chains of spacers can promote protein adsorption [24]. The copolymers of spacers show more protein adsorption than the homopolymers chains [25]. Finally,

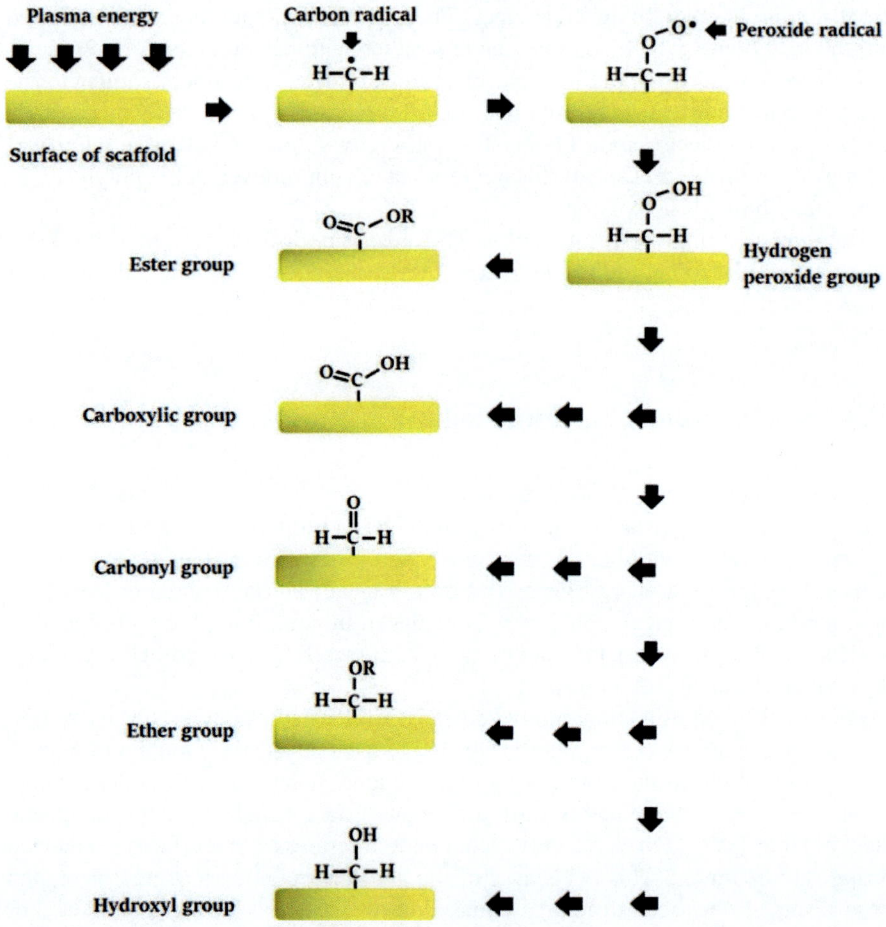


**Fig. 12.3** Example of reaction during 2D scaffold treatment with plasma, or low energy treatment, under $O_2$ condition

some research studies reported that the hydrophilic characteristic has an effect on the promotion of cell adhesion [26–29].

Accordingly, due to the immobilization of spacers on the scaffold surface, some biological molecules, which act as signals to induce tissue regeneration, are conjugated with spacers through a chemical reaction (Fig. 12.3). This can be used to modify scaffolds with biological signals. These 2D scaffolds often have their biological functionality enhanced. Mainly, the scaffolds are modified with biological signals, which are able to induce MSC regulation similar to that of natural cartilage tissue [30, 31]. In so saying, scaffolds treated with reactive chemical groups can then react to biological signals [32]. This shows conjugation, which is generated through covalent bonding, and has high strength. Therefore, biological signals from chemical bonding show a prolonged activity profile [33].

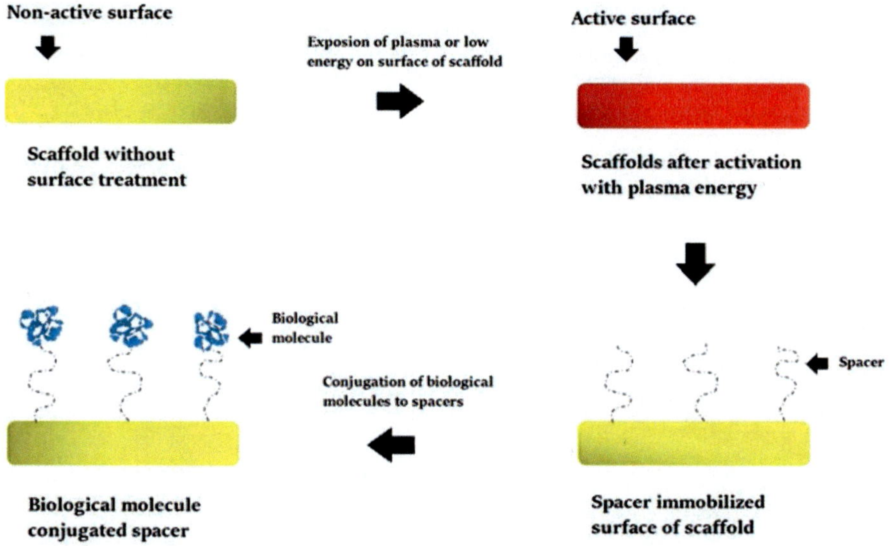


**Fig. 12.4** Immobilization of spacers on the surface of scaffolds, with surface plasma and low energy treatment

The contents in this chapter concentrated on mimicked scaffolds for cartilage tissue engineering. Many studies on these mimicked 2D scaffolds demonstrated that certain structures and functionalities are important for the construction of performance mimicked scaffolds for tissue engineering in cartilage surgery.

## References


1. Vinatier, C., Guicheux, J.: Cartilage tissue engineering: from biomaterials and stem cells to osteoarthritis treatments. Ann. Phys. Rehabil. Med. **59**(3), 139–144 (2016)
2. Grässel, S., Lorenz, J.: Tissue-engineering strategies to repair chondral and osteochondral tissue in osteoarthritis: use of mesenchymal stem cells. Curr. Rheumatol. Rep. **16**(10), 452 (2014)
3. Mollon, B., Kandel, R., Chahal, J., Theodoropoulos, J.: The clinical status of cartilage tissue regeneration in humans. Osteoarthritis Cartilage **21**(12), 1824–1833 (2013)
4. Luyten, F.P., Vanlauwe, J.: Tissue engineering approaches for osteoarthritis. Bone **51**(2), 289–296 (2012)
5. Hattori, K., Ohgushi, H.: Progress of research in osteoarthritis. Tissue engineering therapy for osteoarthritis. Clin. Calcium. **19**(11), 1621–1628 (2009)
6. Smith, B.D., Grande, D.A.: The current state of scaffolds for musculoskeletal regenerative applications. Nat. Rev. Rheumatol. **11**, 213–222 (2015)
7. Girão, A.F., Semitela, A., Pereira, A.L., Completo, A., Marques, P.A.A.P.: Microfabrication of a biomimetic arcade-like electrospun scaffold for cartilage tissue engineering applications. J. Mater. Sci. Mater. Med. **31**, 69 (2020)

8. Chen, W., Chen, S., Morsi, Y., El-Hamshary, H., El-Newhy, M., Fan, C., Mo, X.: Superabsorbent 3D scaffold based on electrospun nanofibers for cartilage tissue engineering. ACS Appl. Mater. Interfaces **8**(37), 24415–24425 (2016)
9. Yilmaz, E.N., Zeugolis, D.I.: Electrospun polymers in cartilage engineering—state of play. Front. Bioeng. Biotechnol. **8**, 77 (2020)
10. Suntornnond, R., An, J., Yeong, W.Y., Chua, C.K.: Biodegradable polymeric films and membranes processing and forming for tissue engineering. Macromol. Mater. Eng. **300**, 858–877 (2015)
11. Torres, J.M., Wang, C., Coughlin, E.B., Bishop, J.P., Register, R.A., Riggleman, R.A., Stafford, C.M., Vogt, B.D.: Influence of chain stiffness on thermal and mechanical properties of polymer thin films. Macromol. **44**(22), 9040–9045 (2011)
12. Andjelić, S., Scogna, R.C.: Polymer crystallization rate challenges: the art of chemistry and processing. J. Appl. Polym. Sci. **132**(38), 42066 (2015)
13. Vieira, M.G.A., da Silva, M.A., Santos, L.O., Beppu, M.M.: Natural-based plasticizers and biopolymer films: a review. Eur. Polym. J. **47**(3), 254–263 (2011)
14. Wei, X.F., Linde, E., Hedenqvist, M.S.: Plasticiser loss from plastic or rubber products through diffusion and evaporation. npj Mater. Degrad. **3**(18) (2019)
15. Frone, A.N., Nicolae, C.A., Eremia, M.C., Tofan, V., Ghiurea, M., Chiulan, I., Radu, E., Damian, C.M., Panaitescu, D.M.: Low molecular weight and polymeric modifiers as toughening agents in poly(3-hydroxybutyrate) films. Polym. Basel **12**, 2446 (2020)
16. Coenen, A.M.J., Bernaerts, K.V., Harings, J.A.W., Jockenhoevel, S., Ghazanfari, S.: Elastic materials for tissue engineering applications: natural, synthetic, and hybrid polymers. Acta Biomater. **79**, 60–82 (2018)
17. Lin, G.B., Ma, P.X.: Synthetic biodegradable functional polymers for tissue engineering: a brief review. Sci. China Chem. **57**(4), 490–500 (2014)
18. Ito, Y.: Surface micropatterning to regulate cell functions. Biomaterials **20**, 2333–2342 (1999)
19. Owen, G.R., Jackson, J., Chehroudi, B., Burt, H., Brunette, D.M.: A PLGA membrane controlling cell behaviour for promoting tissue regeneration. Biomaterials **26**, 7447–7456 (2005)
20. Bae, B., Chun, B.H., Kim, D.: Surface characterization of microporous polypropylene membranes modified by plasma treatment. Polymer **42**, 7879–7885 (2001)
21. Fatyeyeva, K., Dahi, A., Chappey, C., Langevin, D., Valleton, J.C., Poncin-Epaillard, F., Marais, S.: Effect of cold plasma treatment on surface properties and gas permeability of polyimide films. RSC Adv. **4**, 31036–31046 (2014)
22. Alves, C.M., Yang, Y., Carnes, D.L., Ong, J.L., Sylvia, V.L., Dean, D.D., Agrawal, C.M., Reis, R.L.: Modulating bone cells response onto starch-based biomaterials by surface plasma treatment and protein adsorption. Biomaterials **28**, 307–315 (2007)
23. Yakup, A.M., şenel, S.N., Gürdal, A.N., Patir, S., Denizli, A.: Invertase immobilized on spacer-arm attached poly(hydroxyethyl methacrylate) membrane: preparation and properties. J. Appl. Polym. Sci. **75**, 1685–1692 (2000)
24. Gombotz, W.R., Guanghui, W., Horbett, T.A., Hoffman, A.S.: Protein adsorption to poly(ethylene oxide) surfaces. J. Biomed. Mater. Res. Part A **25**, 1547–1562 (1991)
25. Ulbricht, M., Yang, H.: Porous polypropylene membranes with different carboxyl polymer brush layers for reversible protein binding via surface-initiated graft copolymerization. Chem. Mater. **17**(10), 2622–2631 (2005)
26. Arima, Y., Iwata, H.: Effect of wettability and surface functional groups on protein adsorption and cell adhesion using well-defined mixed self-assembled monolayers. Biomaterials **28**, 3074–3082 (2007)
27. Faucheux, N., Schweiss, R., Lützow, K., Werner, C., Groth, T.: Self-assembled monolayers with different terminating groups as model substrates for cell adhesion studies. Biomaterials **25**, 2721–2730 (2004)
28. Keselowsky, B.G., Collard, D.M., García, A.J.: Surface chemistry modulates fibronectin conformation and directs integrin binding and specificity to control cell adhesion. J. Biomed. Mater. Res. Part A **66A**, 247–259 (2003)

29. Hersel, U., Dahmen, C., Kessler, H.: RGD modified polymers: biomaterials for stimulated cell adhesion and beyond. Biomaterials **24**, 4385–4415 (2003)
30. Vinatier, C., Mrugala, D., Jorgensen, C., Guicheux, J., Noël, D.: Cartilage engineering: a crucial combination of cells, biomaterials and biofactors. Trend Biotechnol. **27**, 307–314 (2009)
31. Re'em, T., Tsur-Gang, O., Cohen, S.: The effect of immobilized RGD peptide in macroporous alginate scaffolds on TGFβ1-induced chondrogenesis of human mesenchymal stem cells. Biomaterials **31**, 6746–6755 (2010)
32. Guo, B., Lei, B., Li, P., Ma, P.X.: Functionalized scaffolds to enhance tissue. Regen. Biomater. **2**(1), 47–57 (2015)
33. Salvaya, D.M., Shea, L.D.: Inductive tissue engineering with protein and DNA-releasing scaffolds. Mol. BioSyst. **2**(1), 36–48 (2006)